RESULTS OF IGNORANCE

Obed del Toro Lugo

DEL TORO
Books

RESULTS OF IGNORANCE

@2021 OBED DEL TORO FUGO

Printed in the United States of America

ISBN-13: 978-1-7373762-1-7

LCCN: 2021911733

Del Toro Books

Lakeland, Florida

Contents

PREFACE

Since the year 1962 that I arrived at the United States of America, I have seen and lived all the changes and behavior of the citizens, the politics, and even a worrisome number of Christians; when I say Christians, I am not just talking about evangelicals, I am referring to all of whom proclaim to be believers of Christ. To me, these changes have been disturbing since none of them have been positive. On the contrary, they have been very negative, especially the radical course that many politicians have decided and opposite to human values and what has been established by God the Creator of all positive things. Precisely these disturbing facts have brought me to write the details in this book, with the purpose to see if we can analyze with honesty and transparency, the realities that have been ignored and, without a doubt, have brought are delivering bad results. In all certainty, the country is headed straight into failure, which we can already see in many aspects. For me, it is very worrisome that both politicians and citizens as well as a great number of Christians, are blind and deaf to what is currently happening. Instead of repenting from all the bad behavior and the bad decisions they have made, they continue to lead the country to a sure failure. All the political, scientific, and theological information I have written, is not based in speculations, assumptions, or interpretations; it is based on the scientific and biblical information as well as proved facts, one hundred percent. The founded country and the constitution that was created based on what God had established was left behind, and God has been eliminated from everything. The value that most of the politicians and citizens are giving God, is the same value of one who is completely insignificant. Despite not accepting this fact, for the same reasons, we will face much completed issues than the ones we are currently facing; and that will be the end of America, as I explain it with details, in the next undeniable information that follows.

INTRODUCTION

Ignorance has two definitions. The first is lack of knowledge, and the second is, having knowledge but ignoring it. For many people, to be classified as ignorant is an offense and many people use this term to offend, also using the term donkey or idiot. But there is a reality, we all have been ignorant in any given area. For example, a scientist is very intelligent, but in vast majority, they are ignorant in theology, and this will be fully demonstrated with the information that will follow of the truth and lies of the climate change. In the second term, even though we have information, we chose to ignore. For example, we have knowledge that we should not run through a red light, but we ignore the law, and we run through. In both cases the lack of knowledge or choosing to ignore can bring catastrophic and very bad results. Going through a red light can result in a fine, a death or a death to others. Platon said to Socrates, there is only one good called knowledge; and only one bad called ignorance. In the time we have been living here in the United States, the worldwide country leader and apparently one of the most civilized, we are facing both terms of ignorance in a grand number of citizens. The acts of violence are also due to the lack of knowledge or ignoring the laws. In many of the cases, if not in all, people reach conclusions without having a good knowledge of the facts; and as a result, they create destructions and take the lives of innocent people. Aristoteles said that the ignorant affirm and take for granted, while the wise doubt and reflect. A famous example was the case of George Zimmerman. In the community where George Zimmerman lived, the residents were used to do neighborhood look out, due to the weekly robberies that were occurring, as per the community reports. On this day, Mr. Zimmerman was on surveillance when he spotted TM, a young man walking through the rear parts of the houses. Since that was not the place to walk, and due to what was happening in the neighborhood, he took it as suspicious, and he immediately called the police. He informed them of the suspect, and sated he was following him so that he would not escape. The police instructed him not to follow the suspect and to return to his car; they would arrive within minutes. Zimmerman immediately did as he was told by the police. The police asked him to provide the suspect's description, who was it about, and the color and clothes the suspect was wearing. Mr. Zimmerman told him it was a young black man that was wearing a hoodie that covered his head. A television reporter, with bad intentions, took the conversation out of context, and informed the viewer's incorrectly stating Mr. Zimmerman had killed the young man because he was black. He aired the part of the recording where Mr. Zimmerman was describing the suspect as a young black man but left out the part of the police requesting the description. He then proceeded to emphasize that he took him as a thief because he was a black man that use hoodies to cover their heads. Due to this misleading information, many of whom did not have the knowledge of the full reality, came to their conclusions, and made decisions with billions of dollars in destructions. In the hearings, it was made known by the same prosecutors that prosecuted him, that the case was not based on racism. President Obama ordered Eric Holder to do a civil investigation, to make sure he didn't kill him due to racism. The same thing happened with the cases in Los Angeles with King, and the cases of St. Louis and

Baltimore. The destructions that are currently undergoing due to the G Floyd case, are nothing but ignorant acts due to the following reasons. First, I want to make it clear, that what this policeman did has no name; in any society, a crime like this is not allowed. It is imperative that true justice is made in this case. Nevertheless, without having the evidence that this crime was committed because of the skin color, people have destroyed the country with trillions of dollars' worth in destruction and have mutilated and killed several officers. Not only have they killed police officers, but they have also killed innocent people including children. Since the beginning of the year 2020, up to the month of June 2020, there have been 27 police officers killed. All these things occur because the manifesters believe that through this type of behavior, they will put an end to racism, even though up to this moment, there is no proof it the officer killed Floyd because of his skin color. People are assuming that was the reason. Even if that were the case, my way of viewing things, they are greatly mistaking if they think through that type of behavior, they will end racism. They might accomplish certain things like the police Reform. The Reform of the police can help avoid crimes and abuse on behalf of the police, but it will not end racism. Take note that there is a contradiction between what the protestors claim and the intention of the Reform. The protestors claim that the police kill colored people because of racism. The Reform is to correct the bad application that the police officers use causing abuse and death. What this state is that they have killed not due to racism, but because of bad application of methods such as suffocation or strangulation. Due to these bad applications, year to date up to June 2020, they have not only killed 88 black people, but they have also killed 172 white people, 57 Hispanics and 14 people of other races. This makes it very clear that it is not a matter of skin color. What needs to be understood is the following, no Reform can correct racism. Racism is a bad attribute in the mind, the heart, and the soul. That is why no Reform can detain it. The only one capable of changing this is God and in some cases the education. Like I mentioned, the Reform can correct some things, but it will never correct the bad decisions due to racism. The base of my information comes from the bible, that not even the destructive protests nor the Reforms can stop the abuse and the oppressions, all of this will get worse. The following information makes it clear, that no one can fix this, only Christ will be able to stop the oppression when He establishes his Kingdom here on earth.

Isaiah 11:1-10

A shoot will come up from the stump of Jesse; from his roots a Branch will bear fruit.

2 The Spirit of the Lord will rest on him— the Spirit of wisdom and of understanding,

the Spirit of counsel and of might, the Spirit of the knowledge and fear of the Lord—

3 and he will delight in the fear of the Lord. He will not judge by what he sees with his eyes or decide by what he hears with his ears;4 but with righteousness he will judge the needy, with justice he will give decisions for the poor of the earth. He will strike the earth with the rod of his mouth; with the breath of his lips, he will slay the wicked.

5 Righteousness will be his belt and faithfulness the sash around his waist. 6 The wolf will live with the lamb, the leopard will lie down with the goat, the calf and the lion and the yearling[a] together; and a little child will lead them. 7 The cow will feed with the bear, their young will lie down together, and the lion will eat straw like the ox.8 The infant will play near the cobra's den, and the young child will put its hand into the viper's nest. 9 They will neither harm nor destroy on all my holy mountain, for the earth will be filled with the knowledge of the Lord as the waters cover the sea. 10 In that day the Root of Jesse will stand as a banner for the peoples; the nations will rally to him, and his resting place will be glorious.

Bad behavior has existed since the beginning of humanity. Regarding racism, if we analyze the following, there is no reason to be racist, that is if you truly believe what the Bible says. Besides being a sin, the story states after the flood only Noah, his wife and three kids were left. Genesis 10:1, *This is the account of Shem, Ham and Japheth, Noah's sons, who themselves had sons after the flood.* Verse 32 of the same chapter states that his descendants dispersed through all the earth. This means all of humanity is descendants of the three sons of Noah. Further along I will give the full information in the immigrant's chapter. I mention this because there exists only one reality; even though some are white, some are black, some are brown and yellow, we all proceed from the same descendants. Racism has always existed, but in the times, we are living, politicians and the media are using it as a tool to create hatred and division between humanity for political reasons. I think it is very worrisome to see how a great number of citizens are being prey of bad information on behalf of politicians, and many communication outlets, and from those that are purposely misinforming, inventing, lying, speculating, and in many cases not informing what isn't convenient to them. My way of seeing things, I believe them to be the responsible for all the hatred, divisions, destruction, and crimes. Therefore, it is very important to know what source we are receiving our information from. When we consume food not approved by the Department of Health, maybe with bacteria and venoms, the results are mortal diseases. A medical science book says that there are foods not approved by the Department of Health, such as some herbs, certain seafood, and some other things people eat. These nutrients begin to ruin neurons and as a result you see cerebral and mortal illness. The same thing happens with the current political food. In regards to me, based on the information many people provide through social media, and after taking some time to get correctly informed, for example in the immigration cases, I seek the information from DHLS; and I inform myself by watching the live debates from the Congress and Senate, and other means like C-SPAN News that only share governmental information and accept live calls of the public opinions as well as browsing in http://www.gov.org/ ; I have clearly seen the poisoned food people are taking in on a daily basis from the liberal communications. The results are also catastrophic and mortal. Just as the Bible states, people call evil good, and good evil. United States of America (morally) already died; I don't think it is necessary to give the indisputable reasons; what needs to be analyzed with all honesty is: who have taken them to this death? Daniel 12:10 says: *Many will be purified, made spotless and refined, but the wicked will continue to be wicked. None of the wicked will understand, but those who are wise will understand.* Why do I give this information? Without a doubt, the liberal press is poisoning every day the minds of people with incorrect and ill intended information, and many citizens, including many Christians and ministers are given for certain that information. I want to make clear that the following information is not about defending President Trump, what I am defending is my Christian faith. I do not receive any benefits from Trump or any other political party. I know that the maximum time that any president can remain in his term is 8 years. What the liberal press did to Donald Trump was precisely what they did to President

8

Reagan and both Bush. It is the same thing they will do to any conservative that results elected. On the other hand, all the democratic presidents have their full support. The impartial press no longer exists. What we have for press and commentaries are political activists. The truth of the matter is that this is not about the President's personality, but rather about their political postures. Those that have not noticed this reality must be delirious.

It's important to understand that for the last several years of King Salomon's reign, who did wrong before Go's eyes, and who began to go against God's people, divisions commenced. Through Salomon's son, Israel was divided even to this day. Since the beginning of humanity, the cities, and their leaders, started to serve other gods and doing wrong before God's eyes. Consequently, Israel went into captivity twice, in Asia and in Egypt. Since that moment and still today, God has delivered them, and they have won every battle. The most recent was in 1967. Because they are God's chosen people, and for not converting into Muslims, all the countries of the Middle East hate them. All these countries from the Middle East are the ones who have rebelled against God and are all Muslims. My information is the following: the devil came to steal, kill, and destroy. The devil is currently using politicians that are not only enemies of the United States of America, but of God. The war between President Trump and the politicians in Washington, and the racist accusations, divider, liar, and even dictator, are just mere intents of how to make him fail and remove him from the power. The truth is that it is not about racism, but about ideology and political power. There are a great number of people such as Ben Carson and many others of the black race that know the president personally and assure that he is not a racist. A racist doesn't employee a black nana to work at his home, with the upbringing of their children like he did. All the accusations were political schemes based on speculations, assumptions, and ill intended interpretations. A clear example is the following: In the eight years that Obama was in the presidency, what did he do for the Afro-Americans? Absolutely nothing. When he took the presidency, the poverty level was at 12.2%, when he finished the presidency, it was at 14.8%. The poverty level of the Afro-Americans was left at 26%, and the Hispanics at 23.7%. As for unemployment, the Afro-Americans it was a 11.51% and the Hispanics at 9.3%. President Trump went into the presidency, and through the cut back of taxes and the removal of over two thousand regulations of companies, he created over 7 million jobs. He dropped the national poverty level to 10.5%, the lowest in history. The poverty in the Afro-American community dropped from the 26% to 18% and for Hispanics, it dropped from 23% to 15%. He dropped unemployment of the Afro-Americans from 11.51% to 5.4%, and for Hispanics from the 9.3% to 4.1%. In both cases, it was the lowest in history. Another thing that he did for minorities was, create a judicial reform that included vocational studies. As a result, there were over four thousand minorities were free and working in the fields they studied. What does the liberal press inform? That he is bad, a racist, a divider, despite the results are different.

As we all know, almost 100% of the times in Hollywood movies, we are wrong about who is good and who is bad. At the end of the movie, we cannot believe that those we thought were good, were the bad ones. In Washington DC, a movie started with the title Russian Collusion, the same night that Trump was declared the winner. In the first chapter that took over two years and 45 million dollars, he was declared innocent. In the second chapter of the Pre, Pro, Quo, the calling to Ukraine was also declared innocent. In Hollywood movies, the intention of the promotors is to make the bad be perceived as the good, and the good, bad. In the Washington DC movie, the promoters titled liberal press, made sure to set Donald Trump as the bad one; a divider, racist, female abuser, and a long list of offenses that we all know. On the other hand, the good ones in the movie were the democrats leaded by B. Obama, J. Biden, A. Schiff, J. Nadler, E. Swalwell, and the rest of the gang. As I have mentioned previously in my announcements, I am clear that I don't believe in anything that anyone informs, and even so more when it's information from declared enemies, by word of mouth or written; I only believe in facts that can speak to the truth. Even though many have their minds captivated, consequence of the ill intended information being exposed, facts are proving who the true villains in Washing are. For example, in the first chapter of the Russian Collusion accusation, it was H. Clinton of the Collusion and not Trump, and the evidence was found in the money that was paid so that the accusations could be made against President Trump. In the second chapter of the Pre, Pro, Quo, it was Biden the Pre, Pro, Quo and not Trump. The evidence of the Biden video which showed that if the prosecutor that was investigating his son was not fired, he wouldn't give the one and a half billion dollars assigned to them. They had six hours to do it. Of course, before the six hours were up, he was fired; and of course, without the permission of Obama, he would have never made that declaration. Regarding his son's investigation, he has alleged to have no knowledge, and not taking one cent of the many millions of dollars that his son has received from China and Ukraine. But the emails and his son's former business partner state the contrary. In addition, we know that without his involvement, his son would not have made it to the airport to take the plane and even less earning all those millions without having any knowledge or experience. As for Swalwell, he vigorously accused Trump as guilty in the Russian Collusion case, but now results that he has ties with various enemies of the USA.

Obama declared that Trump was a racist because he had the illegals locked up in cages that he fabricated himself and that he also had them locked up. He liberated them due to a judge's order that told him he couldn't have them locked up for an undefined period. Also, up to this date, he was the president that most people had deported, including Trump. The promoters have taken charge based on their proper interpretations, speculations, and lies, to present Trump as the villain of the movie. Without a doubt, He is a man with a strong, firm character, and this is the reason of his success not only in the presidency, but in his businesses as well. I truly believe this was what the nation needed to achieve, without discussion, what it achieved. Being a president of law and order without being a cry baby, doesn't qualify you as a bad man. Nor because Biden presents himself and talks like a lamb, do we have to classify him as good. This recalls what we see in the book of Revelations 13:11 that states: he has horns like a lamb but talks like a dragon. The following are the facts. Racism; 1); in 1977 Biden said he opposed to the racial segregation because he didn't want his kids to grow up in a racial jungle. 2) As opposed to the Latin community, the Afro-American community has a culture a lot more inferior in comparison to the Latin community. 3) In Delaware, the fastest growing population is of the American Indians that move from India. You cannot visit a 7 Eleven or a Dunkin Donuts unless you have a slight Indian accent. 4) If black people don't vote for me, they are not black. 5) The same person he chose for Vice President, K. Harris, in one of the nomination debates, accused him of racism because in the 70's he had a good relationship with segregation lawyers and collaborated with them to oppose the transportation of children to school, while it was being attempted to correct segregation. Regarding the matter of abortion, which 60 million of babies have been executed, the bad man Trump was opposed, and the good man Biden agrees and voted in favor. Marriage of the same sex, Trump is against it and Biden is in favor. Birth control pills for minor girls, Trump opposed, and Biden was in favor. Stopping prayer in schools and prohibiting to talk about God, Biden voted in favor and have it been forbidden. Trump eliminated the prohibition and brought back prayer and ordered back the ability to talk about God freely. The promoters talked very poorly about Trump, and they accused him of being a criminal because he ordered the death of the Iran General terrorist, who was organizing an attack against the US embassy in Iraq and was also responsible to having killed more than 600 American soldiers. Never once was mentioned that during the Obama-Biden administration, they ordered 503 of the same attacks against the ISIS and Al Qaeda terrorists, and in one of those attacks, they killed 55 civilians including 21 children, 10 women, 5 of them pregnant. In August of 1996, the Democratic leader Biden, worked vigorously against the elimination of the 936 in Puerto Rico, the one Bill Clinton signed, which has brought financial disasters to the Island, and left hundreds of thousands without a job. In 2015, the Puerto Rico government request economic support to Obama-Biden, for the same disaster that Biden caused on the Island, and the help they provided was, impose a delegation to control the costs on the Island. In 2019, President Trump assigned $13 billion dollars of economic relief to Puerto Rico and according to the reports of the Orlando Sentinel and sources in the Island, Puerto Rico is being benefited. He also achieved agreements so that the Pharmaceutical Industry that had left the Island because of Biden, return, and create thousands of new jobs and a healthy financial state for Puerto Rico. There are many comparisons I could mention to reveal who is good, and who is bad, according to the facts. I only mention a few. Appearances, the promoters, and the liberals deceive, facts are what truly talk. The movie is coming to an end, but as in every movie, the end is very dramatic. We will see what the end brings, which does not seem to promise a victory to the real bad people since in the speech Trump spoke to the manifestos, there is no evidence, none, of him saying anything inappropriate, quite contrary, he said our voices needed to be heard patriotically. In addition, according to new information from the FBI, there are some arrests of ANTIFA with charges of planning the attack way before the manifestation. The end won't be until John Durham finalizes the investigation of how the lie started and was invented about Russia to have Trump removed from the power. You might ask yourself: Why talk about Trump if he is already out? It is just to point out what we will be facing, because it is not about Trump, this will continue against any conservatism. They did it to Regan, to both Bush, and now Trump; and they will continue to do it to every conservative that goes against them. I am warning the church, to get well prepared because under the Biden administration, we will be greatly persecuted, and this should teach us the importance of who we support. What I can guarantee is that this is far from over and as in Hollywood movies, we will have Trump part two.

Why do I mention Israel and the Muslims? Because President Trump has become the protector of Israel. As a result, these two Muslim congressmen R. Tlaib and A. Omar, have been attacking and making racist comments against the Jew and the American System. The President is not perfect like no other man is, and I confident to say that not many can withhold the infernal attacks that he has been subject to. It is very sad to see many Christians including many ministers, attacking and accusing him of racist, without any evidence other than those mentioned by his political enemies. The responsibility of any citizen, and more so any Christian, is not to use social media to spread hate and rebellion against any president, or democrat, or republican. What they need to do is as the word of God says in Lucas 6:27-37:

"But to you who are listening I say: Love your enemies, do good to those who hate you, bless those who curse you, pray for those who mistreat you. If someone slaps you on one cheek, turn to them the other also. If someone takes your coat, do not withhold your shirt from them. Give to everyone who asks you, and if anyone takes what belongs to you, do not demand it back. Do to others as you would have them do to you.

"If you love those who love you, what credit is that to you? Even sinners love those who love them. And if you do good to those who are good to you, what credit is that to you? Even sinners do that. And if you lend to those from whom you expect repayment, what credit is that to you? Even sinners lend to sinners, expecting to be repaid in full. But love your enemies, do good to them, and lend to them without expecting to get anything back. Then your reward will be great, and you will be children of the Most High, because he is kind to the ungrateful and wicked. Be merciful, just as your Father is merciful.

"Do not judge, and you will not be judged. Do not condemn, and you will not be condemned. Forgive, and you will be forgiven."

People who feel resentment and hate against a president from any political party, should read what the word of God tells us in the book of Romans 13:1-7, and 1st Peter 2:13-14, *Submit yourselves for the Lord's sake to every human authority: whether to the emperor, as the supreme authority, 14 or to governors, who are sent by him to punish those who do wrong and to commend those who do right.* I believe we must follow the example of Billy Graham's son, and a vast number of Pastors from God's Assembly Church and other organizations. They have informed to be aware of the infernal attack against the President since the night he was elected, and they have created a national prayer chain so God will protect and illuminate him. Pastor J. Hagee, that without a doubt, is a man of God, appeared in a TV show on 7/14/2019 and explained he knew the president personally, and expressed the importance of having him reelected and not let any of these liberal extremists take the power because the results would be catastrophic for the country and the church. His words are being accomplished to perfection.

As for the politicians, they are the real responsible for the violence here in the USA. If we educate ourselves, since the 1850's, various violence attacks were registered due to different ideology reasons as well as for racism, such as the KKK cases. There have been attacks that took the life of President Kennedy, the illustrious Rev. Martin Luther King, and complots to kill others like President Regan and Obama. Nevertheless, since 1995 with the Timothy McVeigh case until today, the attacks against citizens have been gigantically increasing. In general terms we are in the middle of a sick society, as well as a worrisome number of people with a diabolic mind. This society has been created the same legislators in Washington D.C., by taking away the power of the parents to correct their children and removing God from everything with the excuse of separation of church and state. Now a day everything is abuse. Don't get me wrong, I do not approve any type of abuse. But, even reprehending them and not giving them permission to be at places we know are not convenient, is considered abuse. Even speaking strongly to them is reported as psychological abuse. Democrats today have been approving and supporting everything that goes against God and giving it as acceptable, because everything that God established is bad for them. What results can we expect? For the last 40 years until today, the society they have raised, has been growing and they are now, by political convenience, looking for who to blame, when they themselves are to blame. This is the same thing as those parents that let their kids do anything they want, and do not discipline them. At the end and generally, they are the wicked in society. An example of this is of the man that through an assault method, killed another man. He was sentenced with a capital punishment. The day that he was going to be executed, his last desire was to see his mother. When his mother embraced him, instead of loving on her, he tried to strangle her. When they questioned his actions, he answered: "If you would have corrected me when I was a kid and brought home stolen stuff, they would not be executing me today." In this information of Results of Ignorance, I speak on subjects such as: the lies and truth about climate change, immigrants, Christians and politics, theological information that contradicts other subjects such as homosexuality, abortion, and the abuse to social programs. I also quote biblical information in reference to the controversial subject of the war in Iraq. These are controversial subjects that many do not like to discuss since most people, including religious leaders, live intimidated by other's opinions and the laws that have already been legislated. Nevertheless, therefore I believe that the modern world is currently living like Sodom and Gomorra, in an abyss of doom.

It is very clear the great number of people that have lost their principles and morals. As a result, we are living in the middle of a sick society, who has lost its shame, manners, honesty, tolerance, and does not demonstrate to have love for their neighbor, and even less fear of God.

The information I share is both political, as well as theological and scientific. It should be analyzed with honesty and transparency since everything I share is based on personal experiences and my theological and scientific knowledge. The rest of the matters I speak about, I have studied very closely, and I have confirmed them in a neutral state of mind, like the following subject on the climatic change.

THE LIES AND TRUTH OF CLIMATIC CHANGE

Every human being, no matter how genius they may be, even without a first-grade education, are gifted in something. There are many that have not studied art, but they have an impressive talent to paint. Same thing happens in music and every other matter. In the case of the politicians, without a doubt, the same thing occurs. There are many politicians with a precedent ability and an extraordinary talent to convince their followers in believing everything they say and propose.

In this information of scientists and theology, we can clearly see how politicians are deceiving citizens with the climate change matters. It is very worrisome to see how the country has changed in the last 50 years, especially in the last two decades. It is very worrisome to see how the politicians are bringing the country to a bankruptcy, but even worse for me, is the amount of people drugged and believing blindly everything they inform and propose. I think it's time for all American citizens, to stop following politicians that offer false wrong promises and start to analyze with intelligence and honesty, the truth of what's happening in this country. For the lack of knowledge, not only people perish, but entire countries suffer bad consequences, like in the case of Cuba that amazingly has been for over 61 years, submerged in misery. Another country is Venezuela that has the same situation and same path. For what we can see clearly, this country seems to be headed towards the same path. One clear example is the following: its regarding how some politicians are using scare tactics on the citizens with the lie of the climate change, just to obtain power. What follows is the scientific and theological information that will set clear, that the climatic change in context, is just a big lie they want to set.

What is the truth regarding the climate change? The truth is, since the planet was created, it has always suffered changes in climate. We will see this information further along in the data that the scientists give. What is the lie of those that promote fear to the climatic change? The lie is in the context in which the politicians, for political reasons want to impose it.

Some scientists and politicians believe in the global warming based on a theory. They also have a theory that all the planets and universe formed from a great explosion known as The Big Ban Explosion. But not all the scientists agree with this theory of global warming solely based on human activity. There are not only many other factors for the climatic change, but there is also clear evidence that the changes in temperatures, have always existed. There is evidence of hundreds of years ago that the temperatures were much higher than what we currently see. The National Oceanic and Atmospheric Administration (NOAA) states that over 56 million years ago, the average planet temperature was 73 degrees Fahrenheit, which represents 14 degrees higher than what we currently have of 58.71 degrees Fahrenheit.

The following information that I will share is from other scientists as well as what the Bible teaches us about the past, the present and the future of this planet. The word theory plays a big part in this because a theory is not concrete. The information of the theory has its roots in the science books.

<u>The scientific information is that there are many factors as to why temperatures rise and fall.</u>

(Quote) Climate is the condition of the atmosphere in a determined moment or during a short period of time. On the other hand, very often it is described as average or the habitual climate that an area experiments, during a long period of time. This definition of climate is fixed and predictable. But this is a false supposition because any period used to evaluate the average climates, can give abnormal results. <u>For example, many parts of the world experimented medium temperatures, considerably higher in the period of 1945-60, than what they had for 100 years.</u>

Factors that influence in climate: diverse factors stop the regional climates from occurring in simple latitudinal bands. One of these factors is the nature of the earth. The mountains, for example, have a considerable influence in the climate because the serve as hurdles to the wind and because the temperatures go down with the altitude increase in approximately 3% F to every 1.000 feet high: (6,5 centimeters for every 1.000 meters). Many mountains are rainy, while others relatively dry. The higher mountains also affect the air movements in the superior atmosphere. For example, the current that flows towards the west goes up towards the north, over the Rocky Mountains, but turns towards the south on the other side. The effect is to maintain relatively warm air over the Rocky Mountains in a higher level.

The configuration of the continents and the proximity to the ocean, are also important because the great water extensions (including the lakes) tend to moderate the climate. The coastal lines of the oceans and lakes tend to have climates less extreme, than those places in the center of the continents. This moderate influence of the water is greater near the ocean, that not only retains the heat easier in the earth, but it also transports it. That is why the ocean currents, both warm and cold, play an important role in the determination of the climates of the coasts.

<u>The development of the air masses, continental and maritime, is another factor that influences dramatically the climate.</u> This is exemplified when the changes in station provoke reversions in the air's direction or monsoons. The monsoonal climate is more evident in the south of Asia, where the quick change to cold weather in the winter, provokes an air mass development of high pressure over the earth. Of this air mass, dry winds will blow over the northeast. In spring, the movement of the sun towards the north, makes the northern part of India to heat up and as a result, a low-pressure system develops, and the adjacent winds of the southeast are absorbed through Ecuador, changing direction, as they do, to humidify the west direction.

Local Climatic Factors

The local climates are influenced by special factors that operate in comparatively small areas, for example, diverse local winds. The Fohn winds of the northern foothills of the Alpes, blow when the low-pressure systems on the north of Europe absorb the winds of the south. As they descend, the Fohn winds warm up causing rapid temperature increases in the areas they pass by. The Chinook, a similar wind that forms at the end of the winter and the beginning of spring at the east of the Rockies, can elevate the air temperature to 77 degrees Fahrenheit (25 C) in three hours.

Another local influence is the solar radiation percent reflected by the surface or albedo. The recent fallen snow has an albedo of about 9 percent, which explains why it doesn't melt with the glowing solar light. The dry, sandy grounds, have a higher albedo than the dark, clayey ones. The forests have lower albedos, but the forest grounds tend to stay cool even in the warm days, because a great part of the sun radiation is absorbed through the trees and very little penetrates at floor level.

Climate Classification

Climate has a great influence on the ground and vegetation, but the climatic regions, like the ground and vegetation zones, rarely have precise boundaries on the ground. In change, a climatic type, generally merges imperceptible in another. There have been several intents to produce climatic classifications worldwide. The ones used the most were from the Soviet Scientist Vladimir Koppen, who between 1900 and 1936, published a series of classifications of different complex grades. Essentially, he tried to relate the climatic characteristics and the vegetation, utilizing two bases. He divided the world into five designated principal regions: A, B, C, D and E.

Changing Climates

The existence of carbon seams in the Antarctic and dinosaur fossils in Spitsbergen (that is in the artic polar circle) show that the climate has changed dramatically during the millions of years of earth's history. We also know that the position of the continents has changed, and are still changing, as result of the earth's plaque movements. So, we can postulate, for example, that during the cretaceous period (that lasted from 130 to 65 million of years), when the fossil evidence showed the aged, luxurious fig trees and ferns grew in the Disko Island in Greenlandic. This island must have been indispensably closer to the Ecuador than how it is today. But the earth plaque movements are now slower, averaging a bit more than a centimeter per year. From there the advances and retreats of the large layers of ice during the recent Ice Age of the of Pleistocene (of approximately 1.800.000 to 11.000 years) and the climate fluctuations more recently experimented in the last 100 years, cannot be explained by the tectonic plaques.

Evidence of the climate fluctuations

The evidence has been accumulating frequent climate cycles, with warm periods or alternate humidity, cold and dry. During the ice age of Pleistocene, for example, there were 6 to 20 periods important in Europe when the ice advanced, and these glacial ages were punctuated by interglacial phases (also called interstitial), even though the date has not been predicted for the sixth glacial aid. The evidence comes from different fountains, including drilled nucleus rocks from the marine beds. In these basic samples the fossil abundance of certain marine organisms that proliferate during warm conditions and they turn scarcer during cold periods and show cyclical variations, indicating that the climate also varied periodically. More tests have been obtained from analysis of ice nucleus in the ice layers, ground samples and tree rings.

As per recent findings, it appears the north hemisphere had a warmer climate between 900 and 1300 D.C. than now. It was in the X century, that the Nordics founded a settlement in Greenlandic, where the average temperatures were estimated between 1-7 F (4-C) higher than today, but this settlement had disappeared at the end of the XV century probably due to the climate worsening. In Europe, the period of 1450-1850 is often called the Small Ice Age.

The causes of climate change

The causes of the climate fluctuations have not been cleared completely, although many theories have been suggested. Some scientists believe that the small variations in the earths' orbit around the sun, are the main cause. Others have assumed that the miniscule alterations in earth's inclination in it's axis could provoke the weather belts change, altering the climate. It has also been suggested that the fluctuations through the short and long term of the sun's activity, the stains caused by the solar cycles every 11 to 17 years, could have affected the climate. Changes could also have occurred after a prolonged volcano activity.

The volcanic dust could reduce the amount of solar radiation that reaches the surface, causing changes in the climate. After the Krakatoa volcano in 1883, for example, the dust remained on the surface for over a year. During this time a 10% solar radiation drop was registered in southern France. There is also a concern that the important climate changes could have resulted from human activity, such as deforestation and the atmosphere contamination.

World climate is influenced by many factors, including the atmosphere contamination. Some scientists have postulated that an increase in air pollution will provoke the temperatures worldwide to increase, because the pollutants help retain earth's warmth. Others believe the main effect of the air pollution is blocking the sun's rays, resulting in temperature decreases. Nevertheless, neither of these predictions have been confirmed.

The following information comes from a scientist.

Earth has approximately 4.6000 billion years, and during this vast amount of time, it has had (seven big glacial eras), that should not be confused with glaciations. Despite of this, for the most part of our planet's history, the climate has been a lot warmer than the actual. Not in vain, despite the current warm phase, we find ourselves immerged in a grand glacial era (the seventh). A glacial period is a prolonged period, in which global temperatures decrease resulting in the continental ice expansion of the polar casquets and the glaciers. The glaciations subdivide into glacial periods, being Würm the las one until today.

During the first 2.3000 billion years of the planet (half of its age), Earth was a lot warmer than today without ice on the surface.

THE EARTH'S CLIMATE ALONG HISTORY:

After approximately 300 billion years, the planet became warm again without any known cause. Ice began to disappear and the great ocean that covered the earth started to be populated by living organisms, even more complex. That is how things continued to occur until the cold came back to the picture. 1.2000 years ago, approximately, it is believed to have passed the second "White Earth". New life forms suffered although some, the ones that adapted more, resisted in the ocean bottom and in the Equatorial Zone, free of ice. After that second glacial era, a warm season followed, shorter than the one before, since 700 million years the third episode took place (Earth Snowball). Of the four extreme cold phases and huge ice extensions, that is believed to have trespassed our planet throughout history, this third one is believed to be the most important of all, since there seem to be indications that the ice reached all the way to the Equator. The only life forms that survived this episode must have been submarine. That is how things transitioned for 150 million years reaching the Precambrian (since 550 million years), having gone up to that moment 88% of the earth's age. Precambrian means that it is the oldest and it precedes the primary era or the Paleozoic; it extends from the earth's cortex formation 4.500 million years ago up to the ocean's life beginning about 570 million years ago. The Precambrian is characterized mainly by the almost absolute absence of fossils and an intense volcanic activity.

It is believed that the volcanic activity could melt the thick ice layer that developed during that episode of "White Earth", thanks to a potent greenhouse effect that counteracting the heat loss that escaped the frozen surface towards space. From that moment on to this date, the hot cold pattern has not stopped repeating itself, although with different magnitudes and scales, as per the seasons. The climate suffers a new reverse during the finalization of the Carboniferous (carbon mineral) 300 million years ago, progressively getting colder, until the fourth "White Earth" took place, although it is not entirely a known fact which extension the ice reached during this episode of cold in the planetarium scale, or in the previous three. A new radical change is produced in both the climate as well as in the scenery, in difference of what occurred in the other "White Earths", on this occasion the planet is geologically different. 500 million years ago, a great ocean dominated the earth with several big islets. 300 million years ago, those huge earth masses joined and formed the super continent Pangea. In geological seasons after, the super continent starts to fracture until a distribution of oceans are formed and a similar continent is formed like in actuality, 50 million years ago. This circumstance, combined with other internal factors (volcanic activity), and external (astronomic), have an important implication in the climate behavior, since the ocean currents (deep and superficial) are the big modulators of the earth's climate.

Towards the already mentioned year 700, in high altitudes of the northern hemisphere, an exceptional warm season begins that lasts up to the year 1200 approximately and in climatology receives the name of Small Climatical Optimum or Medieval. Currently there is a scientific debate questioning if in this period, the warming was of higher or lower magnitude than the one we are currently living. The transition of hot to cold was characterized of being an extremely humid time, that gave way to colder years to what resulted as the beginning of the Small Ice Age that would last until the XIX century.

In Spain we can fixate the beginning of the small ice age towards the year 1500. Although the small ice age is not comparable in duration or magnitude to a glaciation, it was sufficiently important as to influence in the development in the European Civilization and other parts in the world. The small ice age consisted in general lines in the succession of 150 uninterrupted years with long and very cold winters, and short warm summers, even though during this period, the climate change was not global since some indicators point that the southern hemisphere of the earth barely reflected its effects. Also, we cannot give an official beginning and end date for this period since there are different temporary phases, depending on the affected regions. Nevertheless, the 1500 to 1700 period is considered as the coldest, initiating a cold period in some places at the end of the XIV century and prolonging in others up to the XIX century, with important high-lows during those almost five history centuries. Between 1565 and 1665, the winter sceneries, became a frequent motive between the European painters (Pieter Brueghel, the old man, is one of the best examples), what is very clear evidence of the type of dominant time of that period. The two main causes that presumptuously unchained that cold period in history. The solar activity was one of them. Concretely, the period that runs from 1645 to 1715, the sun had a very abnormal behavior, with barely stains on its surface in what is called Maunder Minimum. <u>This period coincided with the years of very low temperatures of all the small ice age.</u> At the end of the small ice age, the same thing that happened in the beginning occurred, that the climate suffered several high-lows with extremely rainy years like 1846, in which camps in Ireland flooded and the potatoes were ruined. That provoked in the green island the Great Hunger that will prolong to 1850, were up to a million people died due to hunger and sickness, provoking a massive exodus of people from Ireland towards Great Brittan and the USA; another great evidence of the influence that the climate has and will have in history. Climate is always a factor to take into consideration although we shouldn't always establish a cause-and-effect relationship. Nevertheless, there are clear cases, for example, the terrible droughts that occurred in the XX century in the Sahel zone that have enormously conditioned the way of life and the customs of the habitants in countries like Mauritania, Mali, and Senegal. The last and great drought in the zone occurred between 1968 and 1973, taking with it the lives of a quarter of a million people. That vast region, south boarder of the Sahara, far from getting back up, it has become even more deserted and therefore many of its habitants have gone elsewhere to survive. The period that goes from 1850 to our current days, covered completely by the registries of the climatological variables, if we compare to other historical periods that has been very commented, we can consider it a warm and benign period that has contributed without a doubt, to the economic and population growth that has occurred through humanity. Through that time, 162 years, the climate has not behaved in a uniform way, but rather we can distinguish three great periods. The first one would be from 1880 to the decade of 1940, characterized mainly by regaining, slowly but surely, the warmer temperatures. This tendency broke between the decades of 1950 and 1970 to initiate in the 80's of the XX century, a new warm phase, which is where we currently are, and the scientists relate the climatic changes.

Politicians for political reasons, have used the climate change, based on the contamination, not just a reality, but ultimately a crisis as well. The latest information thy are providing is that the world will end in 12 years if we do not do something immediately. As a result, they propose a new agreement called: The New Green Deal, as they say this will avoid a catastrophe. They say they are taking measures to save the planet. The United States is just a small land with 330 million people. My question is: how are they going to control the rest of the world to control the climate change? In second place, the information of other scientists and the information in the Bible are clear regarding who oversees the climate, and the changes in the sun, the moon and the stars, and the distance of the sun from the earth's orbit in an elliptical orbit, and its rotation, that as per the scientists, is the primary reason of the temperature changes of our planet.

(Once again, I quote), in prior geological periods, after the super continent, the islets fractured and ocean distribution was gained, and continents like the current ones appeared 50 million years ago. That circumstance, in combination to other internal factors, (volcanic activity) and external factors (astronomic), has a very important implication in climate behavior, since the marine currents (deep and superficial), are the biggest modulators of the earth climate.

It is very clear that pollution is not the only factor in climate change. What this means is that no matter what politicians say that they will do regarding the climate if they are elected president, it is a big lie and its proof they are only looking to be in the power by betraying the citizens. The following is more concrete evidence.

As per the Science World Encyclopedia, there are scientists that have the (theory), that contamination provokes worldwide temperatures to drop. But the causes of the climatic fluctuations have not been proven yet. There are other scientists that have other theories. Some scientists believe that the small variations on earth's orbit around the sun, affect the intensity of the solar radiation that arrives on earth, and this is the main reason of the global warming.

Many scientists are still studying the climate changes and verifying temperatures, as well as the levels in the ocean. No one can deny that there is nothing concrete yet, without any doubts or clear evidence, that the planet is heating up, only due to human activity. This is the reason why many people, politicians and scientists disagree with the reasons on climate change.

Some politicians have an urgency to promote the fear of global warming that they are misinforming people on television programs, stating that every scientist agrees with the end of the world due to global warming, and this too is another lie.

There is a Scientific Geologist that has been studying the earth for over 35 years. He is the author of the book: Inconvenient Facts, the science that Al Gore does not want you to know. He was on the Fox News Show with Laura Ingraham, on the 12th of March 2019, in which he stated he disagrees with the information of other scientists regarding the information on climate. Another ex-scientist from NASA wrote a book that explains in detail the reasons the politicians in Washington have, and why at all expenses they are pushing a false theory.

March 12th, 2015, the secretary of state of the United States of America, Mr. John Kerry, had a conference in the Atlantic Council in Washington, DC., and specifically poke about climate change. Climate change is the same thing as Global Warming. The change of the name is because with the global planet warming, they have not been able to fully convince people. Besides this, they have not been able to have the support of other scientists as well. This same week an ex-scientist from NASA announced on News Max, and he categorically explained that the global warming or climate change is false. This is the scientist I previously mentioned, that exposed the reasons these big politicians have, to support these theories. With a desperate effort to make a reality the climate change, it is nothing more than a trillion dollar   industry. Precisely this NASA scientist gives this information on his report. Besides the trillion-dollar industry, the information is that the politics that are supporting this theory is with the purpose to impose and collect 22 billion dollars a year in taxes to taxpayers for this cause. Mr. Kerry in his conference, after giving an explanation on how the climate change is affecting the planet, and for example he said, that the fish in the sea are already moving towards the north because of the changes in climate. Also, in his speech he mentioned the same information we have been hearing for years of the ice melting, the changes in the water, etc. After all his ranting the most important thing I heard was about having to invest millions of dollars in specialized technology for this purpose since, as per his information, this is the equivalent of creating job offers in all corners of the earth, and these were his own words. Categorically the secretary of state, Mr. Kerry said that the Clean Energy Program (CEP), is an industry of 17 trillion dollars. These are the reasons why President Obama and the Democrats are pressuring with the topic of the climate change. They have also promised to do whatever is necessary in his last two years of presidency, stop the climate change. They are so desperate to promote the global warming that President Obama used the Pope and brought him to the White House, so that for the first time in history a Pope could be used to push the agenda.

It is very important to state that without cooperation of all the industrialized countries of the world, United States can't do absolutely anything to control the industries that contaminate the air around the world. For example, China is not doing absolutely anything to promote in reference to global warming and China is the country with the most industries and habitants in the world. United States in comparison with China and Russia, in comparison with territory and habitants, is a small country. Also, India and other countries are not doing anything to control the air pollution or at least they have not demonstrated to care and are not showing to be desperate like the politicians in Washington. They too are contaminating the air, far more than the United States.

I believe that for a smart believer, few words are necessary. Besides that, no country in the world can control what belongs to God. The Bible speaks clearly regarding whom this planet and the whole universe, belongs to. Psalm 24: 1-2 states:

> *The earth is the Lord's, and everything in it, the world, and all who live in it; for he founded it on the seas and established it on the waters. Hebrews 1:2 but in these last days he has spoken to us by his son, who appointed heir of all things, and through whom also he made the universe.*

The Bible also clearly states what is going to happen to the planet in a very near future. It not only talks about the planet but of the people that live here as well. The Bible says that this planet will go through a process in which three fourths part of the earth will disappear. It continues to say that God will create a new sky and new earth. Regarding this information, I will speak specifically about what the bible says.

It is also important to mention that the condition the country is already in regarding a $30 trillion dollar debt and growing, it is not able to be investing trillions of dollars in this cause. The own democratic party is the one announcing the trillion-dollar expenses. It should also be known that the democratic party and Obama, were investing behind the back of the citizens, millions, and thousands of dollars for this cause, according to the press. For example, on the 6th of march of 2015, the Washington Beacon informed that the democratic leader, Mr. Harry Reid, steered tens of millions of dollars from Biofuel company a California Corporation. The information talks about 770 million dollars and another 245 million dollars they requested for the Clean Energy Project. This totaled over one thousand million dollars in the year 2013 itself. The information also says that Mr. Reid was accused of giving counsel to businesses in his state of Nevada on how they could receive funds using the law of American Recovery and Reinvestment, better known as the stimulus act.

The two scientists previously mentioned are well known at national level. One of them works at the Florida University and the other worked at NASA for 35 years. This last scientist mentioned is one of the most renowned scientists of the country and is an expert in climate system. He is also very known in the meteorology predictions. These scientists as well as others were very clear that global warming or climate change, is nothing more than a great lie in the context that they want to impose it. This scientist states that in 2007, while he still worked in the government, he found evidence that was hidden by the government, scientific evidence that proves their own arguments wrong, regarding the earth warming up. His information demonstrates that a network of politicians, businessmen, and scientists, that are conspiring to spread fear of global warming. He also informed that part of this lie, that the NASA information shows that the planet's temperature from 1979 to 1998 rose .36 degrees, but after 1998 the earth's temperature has been decreasing and is currently 1.8 colder than what was registered in the 10 years between 1979 and 1998. I believe this is the reason why, for political reasons, they changed the name from Global Warming to Climate Change. It is also true that 20 years have gone by since this information was shared, but a grand number of scientists that continue to study the earth inform that since 1880 to the present date, earth's temperature has increased 1.4 grades.

Nevertheless, the new information reveals that the sun is entering its cycle of changes of Umbra and Penumbras, and it is expected for the earth's temperatures, because of the changes, to decrease dangerously for the next 30 years.

Other scientists report the following: without sounding alarming, the truth is we have entered a new climatic cycle, never known by humans, but familiar to the earth, to which we need to adapt to avoid a human catastrophe of great dimensions. Our adaptation to the climate change will be better or worse depending on how wars are resolved, hunger, social inequality, and a long list of problems that we have on top of the table. This is the challenge that humanity faces in the present century.

As for me, all this information makes it clear that what politicians inform about the end of the planet is fake with the intention of scaring the population and owning the power. What many other networks report about climate change, is nothing else than them taking advantage of the small temperature increase of 1.4 degrees from 1880 to 2018, for lucrative reasons. A scientist classifies the temperature increase to the equivalent of having a thermometer in the house and increasing the temperature from 75 degrees to 77 degrees. It would be a matter of adapting. I don't think you have to be that intelligent to understand that this is a trillion-dollar industry. I do believe that many industries are contaminating the air. I also believe in the clean energy. I am not suggesting that with this information, we do not do our part so that we can have a clean planet and without contamination. It would also be great to see the millions of jobs to appear. That would be good for the world's economy. What I don't agree is, based on that great lie, is that they continue to drown the citizens with more taxes, than what we have, with their purpose of supporting this lie. It's true that the Clean Energy Project can create many job positions with the new technology. It would also create millions of jobs building coils, electric panels, solar panels, electric cars, and many other things such as the machines for everything that has been mentioned. For example, I remember that in 1992, I installed solar panels in my home to heat water. That had a cost of $4,500. With all honesty I was saving 45% of the normal electricity cost. Can you imagine if the government here in the United States imposed a mandatory obligation to install solar panels in every home to heat water? How many millions of solar panels would need to be fabricated?

My prediction is that there will come a day in which they will make every home responsible to produce their own electricity. I visualize in a very near future, a big number of electric cars, but not like the Chevy Volt that must be hooked up to an electric outlet to charge the battery, I visualize them with a solar panel installed somewhere so that the battery can charge. I can also see many homes with solar panels so they can produce individually, their own energy. All this also means the fabrication of machines, transmitters, electronic coil instruments, and many other products. The list is long of the things that must be fabricated. Therefore, that is why Mr. Kerry talks about a 17 trillion-dollar industry and jobs in all corners of the world. All of this is coming. That is the true purpose behind the climate change. It would be something good for employment and the economy and for the clean air, but I think this doesn't belong to the government. I don't believe in any matter, the government should collect from the taxpayers thousands of millions of dollars every year, to create industries for the private sector. This was precisely what President Obama did when he gave $500 million dollars to Solyndra, a company in California that build solar panels and with the taxpayer's money. The industry went into bankruptcy and the $500 million dollars were lost. Obama also gave huge amounts of money to automobile companies so they could build electric cars, and just like Solyndra, the money was lost. All this money lost under the name of Clean Energy Project and the taxpayers. I think the role of the government should be imposing just regulations so that companies can operate in conformity of the established laws. If the democratic party wants to promote the clean energy industry, they should do it by speaking the truth, and not based on a theory that has not been proven yet. I personally believe that the legislators have taken the country to an uncontrollable debt and now they are desperate to see if they can invent something that can produce millions of dollars, since the money they are collecting from the taxpayers isn't enough to pay the interests of the debt.

The earth's temperatures are related to the sun.

The following information is what the book of World Science says about earth's climate relating to the sun. If we analyze it carefully, we can conclude that the information shared by politicians of the global warming being caused only by pollution is a lie.

The information is as follows: Our planet depends almost completely on the light and heat of the sun. The earth also has its own warmth, even though it's very minimal from the volcanos. Without the constant sun radiation, planet earth would be a planet with temperatures below zero degrees. The sun itself has its own variations and as a result there are changes in the energy it releases to the earth. These changes have always been of a concern to scientists because big changes can cause a lot of damage to the earth's climate. The ultraviolet radiation has a variation because the sun has a cycle of 11 years. This cycle also variates from 7 to 17 years. During these hot cycles that produce within the sun and are known as sun stains previously mentioned, have two dimensions known as umbra and penumbra. The umbra is dark, with temperatures of 4,000 degrees F, and penumbra with temperatures of 5,000 to 6,000 degrees F. The earth's temperatures also have to do with the vegetation and the clouds. All the radiation that the earth receives goes back to space, if not the temperature would always be increasing. 34% of the solar light sent to earth is reflected by the clouds and sent back to space. The sun also sends solar waves to the earth, ones stronger than others, but they are absorbed by the atmosphere and clouds. They end up just passing through. They do not last,

they dissipate, and they stop existing. (This is just the nature that God created for man's survival on earth).

Scientific evidence shows that the earth was warmer in past centuries.

According to recent findings, it appears that in the northern hemisphere the climate was warmer between 900 and 1300 A.D. than now. It was during the X century that Normans founded a settlement in Greenland, where average temperatures were estimated 1-7 F (4-C) higher than today. This had disappeared at the end of the XV century probably because the climate worsened gradually. In Europe, the period between 1450 to 1850 was called the small ice age. Although exact figures didn't exist before the invention of meteorological instruments, there is a lot of evidence of historic documents that show the small ice age (including records of the bad harvests and paintings of frozen rivers that never freeze today) and of modern analysis of factors like seeds, and pollen levels on the ground as well as deposits referencing those times. Since 1850, the climate has become warmer, although there has been a slight decrease. Fact is that 1968 the ice reached from the Artic, to the south and to the northeast of Iceland, the first time that occurred in those last 40 years.

There are scientists that believe that both pollution and contamination provoke global temperatures to rise. But the causes of the climatic fluctuations have not been proven yet. Other scientists have different theories. Some scientists believe that the small variations in earth's orbit surrounding the sun affect the intensity of the sun's radiation that hits the earth, and this is the principal cause of earth's warming. But others have planted the hypothesis that the alterations in earth's inclination on its axis can cause the climatic belts to change, therefore also changing the climate on earth. It has also been suggested that the fluctuations in short and long term on the sun's activity, cause solar shadows. This activity caused every 11 years, can affect the climate. Changes can also occur after a prolonged volcanic activity. Volcanic ashes can reduce the solar radiation that falls upon the surface and causing climate changes. For example, after the eruption of Krakatoa in 1883, the ashes remained in the atmosphere during a year. For that time, there was a temperature drop of ten percent in the solar radiation that was registered in the south of France.

There is also a certain concern that the big climate changes can result from human activity such as deforestation and contamination of the atmosphere. Some scientists have stated that the increase in air pollution make the global temperatures to increase since the contaminants help to retain earth's heat. But other scientists believe that the air pollution blocks the sun's rays and as a result temperature eventually decrease.

As we all know, till this day, none of these last two theories have been proven. But the second to last theory mentioned, was the one with which Al Gore became a millionaire endorsing it to be true, without being a scientist himself. This is the same theory that Obama and the democrats are currently using to continue to enforce fear on the population with the purpose of imposing hundreds of thousands of dollars in taxes every year and creating multimillion dollar deals for private companies. They are so desperate, as I previously mentioned, that Obama brought the Pope Francisco, to the White House in Washington, DC, to help promote climatic change as a catastrophic problem for the planet and humanity.

It's incredible to me that even in the presidential debates they use their beliefs to impose fear on the population with incorrect information of hurricane events and the fires in the west side of the country. The information that hurricanes are a result of the climate change and that every year they are bigger and stronger, is false. The following is a list of the most powerful hurricanes that prove their information is false. Hurricane Cuba 1932 with sustained winds of 175 MPH; In 1933 there were also two category 5 hurricanes. In 1969 Hurricane Camille with 175 MPH, Hurricane David in 1979 with 175 MPH, and Hurricane Allen in 1980 with 190 PMH winds, the strongest one registered in history so far. Hurricane Gilbert in 1988 with 185 MPH. Hurricane Wilma was the last big one registered in 2005 with 185 MPH winds. Clearly since they started to show records in 1891, there have always been seasons with many or few hurricanes. For example, the busiest season was in 2005 with 28. In accordance with history and records, we will continue to see the same events every year. If the information the politicians give was correct, we would see this increase every year and we would have already seen a hurricane bigger than Hurricane Allen in 1980. Of course, every time there is an active year, they use it as example to blame the climate change. On the other hand, when there is a low season, they say nothing. Regarding the fires on the west, since the records began in 1850, there have always existed fires that have burned for many miles, destroyed thousands of houses, and many firemen have died. In 1889, a fire destroyed 300 thousand acres, in

1923Berkeley destroyed thousands of acres, 640 structures and 584 houses. In 1933 one of these fires killed 29 firemen. In 1953, there was a fire that killed 15 firemen as well as in 1961 thousands of acres and 484 burned. These last years the information is that the fires have increased due to increase and accumulation of combustibles, wood accumulation, increase of electrical transmitters and power lines. These increases are because of the increase of population in the forests. Even though California has a Mediterranean climate that creates the ideal conditions for fires, and it gets worse with the increase of temperatures, it is nothing new. According to the scientific information, the temperatures have always increased and decreased since the centuries after creation. The only new thing is the increasing population worldwide, space and the carelessness of the humans. We should also understand that the deaths that have occurred during floods and fires are in line with the increase of the population in all corners of the earth. This has also resulted in people building more houses in flooding areas, as well as living in the forest where years ago nothing was built. The world population in 1950 was 2.6 billion, in 2020 the population was at 7.684 billion, an increase of 5,084 billion. Honestly, we should analyze a bit further and more intelligently the information we are given.

What has been confirmed is that the earth temperatures are related to the sun. If we look closely to the press, especially The Weather Channel since they keep records of the temperatures and all types of registries that have to do with cold and warmth, we will see many temperature records from many years ago both high temperatures as well as low temperatures like what we see today. For example, in winter of 2015, it was informed for the first time since they started keeping records, all the surface of the lakes in the northern part of the United States had reached a freezing record of more than 25%

of what was registered in years before. In 2015 it was also recorded the lowest temperatures in many parts of the country. That also included snowstorms that broke the records in the quantity of inches fallen. One of those cities was Boston that broke a record of snow inches in its history. The evidence is clearer than water, we can truly see what President Obama, the Democrats, and some Republicans are trying to do. What I ask myself is where is the intelligence of those that believe in global warming, or in the climate change of the earth regarding pollution? The information the Bible gives is contrary to scientists and politics.

the following information is what the Bible teaches us in reference to the beginning of the universe and is contrary to what scientists believe. The Bible speaks clearly on who is in control of this planet and the whole universe that God created. I don't understand how millions of people say to believe God and what the Bible teaches but believe more in what certain scientists and politicians have to say and for political power influence what goes against the word of God.

In the book of Genesis, in the first chapter it says that in the beginning God created the heavens and earth. Genesis 1,

> *In the beginning God created the heavens and the earth. Now the earth was formless and empty, darkness was over the surface of the deep, and the Spirit of God was hovering over the waters. And God said, "let there be light," And the darkness he called "night." And there was evening, and there was morning the first day. And God said, "let there be a vault between the waters to separate water from water." So, God made the vault and separated the water under the vault from the water above it. And it was so. God called the vault sky. And there was evening, and there was morning the second day. And God said, "let the water*

under the sky be gathered to one place, and let the dry ground appear." And it was so. God called the dry ground land, and the gathered waters he called seas. And God saw that it was good. Then God said, let the land produce vegetation: seed bearing plants and trees on the land that bear fruit with seed in it, according to their various kinds. And it was so. The land produced vegetation: plants bearing seed according to their kinds and trees bearing fruit with seed in it according to their kinds. And God saw that it was good. And there was the evening, and there was morning the third day. And God said, let there be lights in the vault of the sky to separate the day from the night, and let them serve as signs to mark sacred times, and days, and years, and let them be lights in the vault of the sky to give light on the earth. And it was so. God made two great lights, the greater light to govern the day and the lesser light to govern the night. He also made the stars. God set them in the vault of the sky to give light on the earth, it's like governed the day and the night, to separate light from darkness. And God saw that it was good. And there was the evening, and there was morning in the 4th day. And God said, let the water teem with living creatures, and let birds fly above the earth across the bold of the sky. So, God created great creatures of the sea and every living thing with which the water teems and that moves about in it, according to their kinds, and every winged bird according to its kind. And God saw that it was good. God blessed them and said, be fruitful and increase in number and fill the water in the seas, and let the birds increase on earth. And there was evening, and there was morning the 5th day. And God said, let the land produce living creatures according to their kinds: the livestock, the creatures that move along the ground, and the wild animals, each according to its kind. And it was so. God made the wild animals according to their kinds, the livestock according to their kinds, and all the creatures that move along the ground according to their kinds. And God saw that it was good. Then God said, let us make mankind in our image, in our likeness, so that they may rule over the fish in the sea and the birds in the sky, over the livestock and all the wild animals, and over all the creatures that move along the ground. So, God created mankind in his own image, in the image of God he created them; male and female he created them. Bought blessed them and said to them, be fruitful and increase in number; fill the earth and subdue it. Rule over the fish in the sea and the birds in the sky and over every living creature that moves on the ground. Then God said, I give you every seed-bearing plant on the face of the whole earth and every tree that has fruit with seed in it. They will be yours for food. And to all the beasts of the earth and all the birds in the sky and all the creatures that move along the ground everything that has breath of life in it, I give every green plant for food. And it was so. God saw all that he made, and it was very good. And there was evening, and there was morning the 6th day.

It is very interesting to see that the following information from scientists, concurs with what the Bible has established.

Scientists have informed that over 500 million years ago, and ocean dominated all the earth, but that 300 million years ago, a new radical change was produced, in both the climate as well as the earth. Just as there is a difference during other "White Earths," in this occasion the planet is geologically different, with various big islets, and that those huge land masses joined and formed the super continent Pangea. In posterior geological seasons, the supercontinent begins to fracture until a new ocean distribution is formed and other continents similar so the actual. They also informed that after this occurred, then it was that for the very first time in the history of the planet, the life of trees and herbs are formed and eventually the human life. Is this not the same thing that was described in the first chapter of Genesis which was written over 3000 years ago?

That principle according to scientists, that have studied meteorology, geology, and the ocean, is that the earth was formed over 4.5 billion years ago. They themselves are not sure but agree that their studies are estimates. How was the universe formed or how did it come to be as per the scientists? The answer to this question according to the scientists and to the book of science already mentioned is the following. Through the ages scientist have been investigating looking for answers. The first scientists believed that the universe had already existed and had no beginning. It wasn't until the beginning of the 20th century that they have found important information. Of course, all this because they do not want to believe what the Bible teaches. Most of the scientists, at the beginning of the 20th century, came to the conclusion that the stars and all celestial bodies moved, but that were canceled among themselves, creating a static universe which means that they have no movement. But there were big doubts with this belief that the universe was static. It was then that in 1915 a new theory was formed by the famous scientist Albert Einstein, who stated that the universe was relative in degree of severity and with energy that moved. He stated that it was rather in acceleration. After Einstein's theory, they found more scientific studies and confirmed that all celestial bodies move and, until today, it is still the base of all scientific studies.

If scientists believed the information of what the Bible says, they would not have lasted until 1915 to know that all celestial bodies move. Over 3000 years ago in the book of Joshua 10:12-13, Joshua knew that the son and the moon were in movement. Abdias also new that men would travel to the planets and what establish their house there. Airplanes were invented in 1930 but if we analyze what the book of Daniel informs, and chapter 8, he was referring to airplanes when he stated that the male goat traveled to the other side of the earth without touching it. In the prophecy that was given of the ram And the male goat, without a doubt tell the events may be repeated, but identical, up the war with Iraq. The ram, Saddam Hussein and the male goat, George W Bush. He also informed that the Ulai River And this vision given to Daniel was in Babylonia which is now Iraq. Daniel 8:1-7:

In the third year of King Belshazzar's reign, I, Daniel, Had a vision, after the one that had already appeared to me. In my vision I saw myself in the citadel of Susa in the province of Elam; in the vision I was beside the Ulai Canal. I looked up, and there before me was a ram with two horns, standing beside the canal, and the horns were long. One of the horns was longer than the other but it grew up later. I watched the ram as it charged toward the West and the north and the South. No animal could stand against it, and none could rescue from its power. It did as it pleased and became great. As I was thinking about this, suddenly a goat with a prominent horn between its eyes came from the West, crossing the whole earth without touching the ground. It came toward the two-horned ram I had seen standing beside the canal and charged at it in great rage. I saw it attack the ram furiously, striking their ram and shattering its two horns. The ram was powerless to stand against it; the goat knocked it to the ground and trampled on it, and no one could rescue the ram from its power.

Wasn't this what happened to Sedan Hussein with GW Bush?

In reference to the existence of the earth it is only God who created everything, he knows exactly the time of its existence. For me, what is wonderful to see is the perfection with which God made our planet for the life of the human being, the animals, the vegetation, and everything that has life here on planet earth. Most of the scientists have their own conclusions on how the earth was formed, as well as the other planets and the universe. Nevertheless, when we analyze their own information regarding our planet, I can't comprehend why they cannot see how everything was created and choose to believe the theory of The Big Bang explosion that I explain further along.

It is indisputable with the perfection that God made the earth and how it was created with specific purpose, and with the perfect distance from the sun, but even more so with the purpose in the rotation, orbit, and the 90-degree angle that I will explain further along. Referring to the new relativity theory, new studies were formed, and they have concluded that the earth came to be through an unimaginable explosion, that took place between 10 and 20 billion years ago. This is The Big Bang explosion I had mentioned And that it's still being taught in our schools. Scientists say that a grand, condensed, intense nucleus exploded since temperatures we're at billions of degrees of heat. The explosion created an expansion and as a result the universe. The information from the scientists is that immediately after the explosion, temperatures descended in 100 seconds 10 billion degrees. Scientific studies believe that the elements that the explosion created, resulted in The Galaxy, the stars, and the solar system with all its planets. Of course, that includes our planet earth. The scientific information is that every one of them was formed during billions of years. For example, our planet was formed over 4.5 billion years ago. What scientists haven't been able to determine is when was it formed or when the elements that caused the explosion formed.

The planets of the solar system are 9, from mercury to pluton, and they are at different millions and billions of miles from the sun, which is in the center of them all. Mercury is the closest to the sun at 43 million miles. And Pluto is the furthest at 4.5 billion miles away from the sun. Each one of these planets has their own characteristics. For example, mercury due to its poor atmosphere and distance from the sun, has day temperatures of up to 648 degrees Fahrenheit and its nightly temperatures close to negative 315 degrees. Pluto, which is considered not to have an atmosphere, and because of its distance from the sun, has temperatures of negative 369 degrees Fahrenheit. Our planet earth, which is the planet that our God as perfect architect created and designed for human life like I mentioned, is at a perfect distance and has the perfect characteristics for human survival. Our planet is at 94 million, 500,000 miles from the sun in the month of July, and a distance of 91 million 500,000 miles in the month of December. Like I mentioned there is a specific reason why this was created this way. That makes it clear that all this perfection is not due to an explosion.

We should analyze all these facts through God's perfection. Planet earth as well as all the other planets, orbit around the sun. The full orbit of the earth is 365 days. The orbit of our planer is not completely round. It is elliptic. It is elliptic with a specific purpose. This elliptic cause is what allows the earth to be at 94,500 miles from the sun during the month of July and at 91,500 in December. The planet was also created at an inclined position at 90 degrees and while it moves at 18.5mph (30k) in its orbit around the sun, it also shifts. This is what allows the four stations during the year and the change in temperature. These are the details that make everything perfect. Because of its elliptic orbit and its rotation is why summer has a warm beautiful weather and is the farthest from the sun, and for the same reasons its why December has colder temperatures and is the closest to the sun. Without God's perfection we would burn in summer and freeze in winter. Do you really think that all this perfection was created from the Big Bang Theory? Not only this, but God also created it perfectly for the human life, contrary to other planets. Therefore, there are still many scientists studying other planets and celestial bodies to see if they can find possible life on other planets. As I have previously mentioned, with the temperatures in Mercury and Pluto, these planets as well as others, are eliminated from the possibilities of life. Planet Mars in order from the sun is 154,900 million miles away from the sun and even though many scientists are still observing the possibility of it being capable of having life, although its characteristics prove otherwise. For example, our planet has an atmosphere of 9.78 which means it has the force to retain a dense gravity force, while in Mars the atmosphere is 3.72. This is the reason the atmosphere of the planet is too thin, and because of this it cannot retain heat and during the day the temperatures are of -24 degrees Fahrenheit and during the night they descend to -191 degrees Fahrenheit. Our planet earth is also composed of 71% water, an important necessity for life's existence. Planet Mars on the other hand is completely deserted.

God's creation s so perfect and done with an unimaginable power, that the more intelligent a person is, the less they can understand God's greatness. People want to analyze with a human nature, God's nature which are unimaginably powerful. If the Big Bang theory was so unimaginable for many scientists, so much more is the power of God. One of those examples we can see in history books and in the Bible when God instructed Moses and with a simple rod he opened the red sea.

There are many politicians today that seem to believe they are higher than the Creator, who is not only the one to control the earth but the universe as well. That includes everything that's in it no matter what scientists, non-believers and atheist think.

The teachings in the bible instruct us that God is in control of all of nature, the universe, of the hot and the cold, snow, granite, and everything that is under his control. One of the biggest proofs we can see it in the book of Joshua, chapter 10: *Five Kings united to fight against Israel with the intention to take over the land of Gilgal, a very important city in those times. They sent notice to Joshua, Moses's successor of what was to occur.* The story is as follows:

> *Johua 10:7-13*
>
> *7 So Joshua marched up from Gilgal with his entire army, including all the best fighting men. 8 The Lord said to Joshua, "Do not be afraid of them; I have given them into your hand. Not one of them will be able to withstand you."*
>
> *9 After an all-night march from Gilgal, Joshua took them by surprise. 10 The Lord threw them into confusion before Israel, so Joshua and the Israelites defeated them completely at Gibeon. Israel pursued them along the road going up to Beth Horon and cut them down all the way to Azekah and Makkedah. 11 As they fled before Israel on the road down from Beth Horon to Azekah, the Lord hurled large hailstones down on them, and more of them died from the hail than were killed by the swords of the Israelites.*

12 On the day the Lord gave the Amorites over to Israel, Joshua said to the Lord in the presence of Israel: "Sun, stand still over Gibeon, and you, moon, over the Valley of Aijalon."
13 So the sun stood still, and the moon stopped, till the nation avenged itself on[a] its enemies, as it is written in the Book of Jashar.

The sun stopped in the middle of the sky and delayed going down about a full day.

According to scientific information referencing these events, a report stated the following. Occasionally we hear that the NASA computers have proved the unusual day that accompanied the Gibeon battle mentioned in Joshua 10:12-14. This small but spectacular story of the NASA computers started circulating at the end of the 1960's and the beginning of the 1970's during the Apollo program. According to history, when preparing for the landings of Apollo on the moon, a NASA computer calculated the positions of the earth, the moon, and other solar system bodies with great precision that goes beyond past and future. This computer produced a flaw in the fifteenth century before Christ. This flaw was produced because the bodies of the solar system were not aligned in the correct position indicting that almost a full day was missing in time. In addition, a period of 40 minutes was missing various centuries later. This completed a full missing day. The scientists were unconcerned with this fact until one of them said he remembered that when he was little, in bible school they taught him that Joshua instructed the sun not to hide for a full day. That is when they opened the bible to Joshua 10:12-14. According to scientific studies, that hold up of almost a day in Joshua summed up to 23 hours and 20 minutes. This indicated that there were still 40 minutes missing. The 40 minutes were found in 2 Kings 20:8-11 when the Prophet Isaiah Hezekiah had asked Isaiah, "What will be the sign that the Lord will heal me and that I will go up to the temple of the Lord on the third day from now?" Isaiah answered, "This is the Lord's sign to you that the Lord will do what he has promised: Shall the shadow go forward ten steps, or shall it go back ten steps?" "It is a simple matter for the shadow to go forward ten steps," said Hezekiah. "Rather, have it gone back ten steps." Then the prophet Isaiah called on the Lord, and the Lord made the shadow go back the ten steps it had gone down on the stairway of Ahaz.

It is explained as follows. In those times the Acaz clock was used. In this clock 15 degrees are equivalent to one hour, 10 degrees signifies they backed up two thirds of fifteen, which is the same as 40 minutes. The truth of the matter is, I don't know what additional information we need to give people so they can see who has the truth. Colossians 1:15-16:

> **15 *The Son is the image of the invisible God, the firstborn over all creation. 16 For in him all things were created: things in heaven and on earth, visible and invisible, whether thrones or powers or rulers or authorities; all things have been created through him and for him.***

Please take note that it says the visible and invisible. After thousands of years, it is now that the scientists discover other planets and galaxies. Creation is so huge that they will never be capable of seeing everything that exists. God spoke about what can and can't be seen in Psalms 148:1-8:

> ***Praise the Lord.[a]***
>
> ***Praise the Lord from the heavens;***
> ***praise him in the heights above.***

2 Praise him, all his angels;
praise him, all his heavenly hosts.
3 Praise him, sun and moon;
praise him, all you shining stars.
4 Praise him, you highest heavens
and you water above the skies.

5 Let them praise the name of the Lord,
for at his command they were created,
6 and he established them for ever and ever—
he issued a decree that will never pass away.

7 Praise the Lord from the earth,
you great sea creatures and all ocean depths,
8 lightning and hail, snow, and clouds,
stormy winds that do his bidding,

Since everything belongs to him, He is in control of the universe. He has already established a future for the planet and for the people that live in it. It doesn't matter what arrogant politicians believe or what they do with global temperature, they will not be able to do anything.

The following is the future of the planet and the people:

2 Peter 3:10-13

10 But the day of the Lord will come like a thief. The heavens will disappear with a roar; the elements will be destroyed by fire, and the earth and everything done in it will be laid bare.

11 Since everything will be destroyed in this way, what kind of people ought you to be? You ought to live holy and godly lives 12 as you look forward to the day of God and speed its coming. That day will bring about the destruction of the heavens by fire, and the elements will melt in the heat. 13 But in keeping with his promise we are looking forward to a new heaven and a new earth, where righteousness dwells.

Revelations 6:12-13: I watched as he opened the sixth seal. There was a great earthquake. The sun turned black like sackcloth made of goat hair, the whole moon turned blood red, 13 and the stars in the sky fell to earth, as figs drop from a fig tree when shaken by a strong wind.

Zephaniah 3:8, Therefore wait for me," declares the Lord, "for the day I will stand up to testify. I have decided to assemble the nations, to gather the kingdoms and to pour out my wrath on them— all my fierce anger. The whole world will be consumed by the fire of my jealous anger.

Isaiah 65:17: "See, I will create new heavens and a new earth. The former things will not be remembered, nor will they come to mind.

Revelations 8: 6-13, *⁶ Then the seven angels who had the seven trumpets prepared to sound them.*

7 The first angel sounded his trumpet, and there came hail and fire mixed with blood, and it was hurled down on the earth. A third of the earth was burned up, a third of the trees were burned up, and all the green grass was burned up.

8 The second angel sounded his trumpet, and something like a huge mountain, all ablaze, was thrown into the sea. A third of the sea turned into blood, 9 a third of the living creatures in the sea died, and a third of the ships were destroyed.

10 The third angel sounded his trumpet, and a great star, blazing like a torch, fell from the sky on a third of the rivers and on the springs of water— 11 the name of the star is Wormwood.[a] A third of the waters turned bitter, and many people died from the waters that had become bitter.

12 The fourth angel sounded his trumpet, and a third of the sun was struck, a third of the moon, and a third of the stars, so that a third of them turned dark. A third of the day was without light, and also a third of the night.

13 As I watched, I heard an eagle that was flying in midair call out in a loud voice: "Woe! Woe! Woe to the inhabitants of the earth, because of the trumpet blasts about to be sounded by the other three angels!"

Revelations 21:1-6, *Then I saw "a new heaven and a new earth,"[a] for the first heaven and the first earth had passed away, and there was no longer any sea. 2 I saw the Holy City, the new Jerusalem, coming down out of heaven from God, prepared as a bride beautifully dressed for her husband. 3 And I heard a loud voice from the throne saying, "Look! God's dwelling place is now among the people, and he will dwell with them. They will be his people, and God himself will be with them and be their God. 4 'He will wipe every tear from their eyes. There will be no more death'[b] or mourning or crying or pain, for the old order of things has passed away."*

5 He who was seated on the throne said, "I am making everything new!" Then he said, "Write this down, for these words are trustworthy and true."

6 He said to me: "It is done. I am the Alpha and the Omega, the Beginning, and the End. To the thirsty I will give water without cost from the spring of the water of life.

Revelations 21:24-27: [24] *The nations will walk by its light, and the kings of the earth will bring their splendor into it. 25 On no day will its gates ever be shut, for there will be no night there. 26 The glory and honor of the nations will be brought into it. 27 Nothing impure will ever enter it, nor will anyone who does what is shameful or deceitful, but only those whose names are written in the Lamb's book of life.*

The information on the bible is very clear oh who controls the universe. Also, the scientific information is clear on the reasons and why climate and temperatures suffer changes. It is also clear the reason some politicians have that are entering in a territory that only belongs to God. It is clear to me they do it for political reasons without fearing the Creator. Believe it or not what I do inform you is that they will have to answer to God on judgment day without any escape. I not only mention this for all the reasons previously stated, but in addition to all the laws they have created that go against God without any remorse. It is clear to me based on their behavior, how they talk, legislate, and govern, God is a nothing to them. One clear example is abortion. Many democrats have approved abortions up to full term babies and have even suggested they are left to die if born alive. Also, there is the legal marriage of homosexuals approved by the democrats. Everything that is legislated that goes against what God has established has to be called out and contradicted like same sex marriage. The lack of knowledge has many citizens supporting these causes, and others that have the knowledge are choosing to ignore. Both cases are going to be catastrophic, furthermore than the climatic change. We are already experimenting consequences like Covid-19, but many are not aware.

The following is what the bible says about same sex marriage and abortion.

> *Romans 1:18 -32,* [18] *The wrath of God is being revealed from heaven against all the godlessness and wickedness of people, who suppress the truth by their wickedness, 19 since what may be known about God is plain to them, because God has made it plain to them. 20 For since the creation of the world God's invisible qualities—his eternal power and divine nature—have been clearly seen, being understood from what has been made, so that people are without excuse.*

> *21 For although they knew God, they neither glorified him as God nor gave thanks to him, but their thinking became futile, and their foolish hearts were darkened. 22 Although they claimed to be wise, they became fools 23 and exchanged the glory of the immortal God for images made to look like a mortal human being and birds and animals and reptiles.*

> *24 Therefore God gave them over in the sinful desires of their hearts to sexual impurity for the degrading of their bodies with one another. 25 They exchanged the truth about God for a lie and worshiped and served created things rather than the Creator—who is forever praised. Amen.*

> *26 Because of this, God gave them over to shameful lusts. Even their women exchanged natural sexual relations for unnatural ones. 27 In the same way the men also*

abandoned natural relations with women and were inflamed with lust for one another. Men committed shameful acts with other men and received in themselves the due penalty for their error.

28 Furthermore, just as they did not think it worthwhile to retain the knowledge of God, so God gave them over to a depraved mind, so that they do what ought not to be done. 29 They have become filled with every kind of wickedness, evil, greed and depravity. They are full of envy, murder, strife, deceit, and malice. They are gossips, 30 slanderers, God-haters, insolent, arrogant and boastful; they invent ways of doing evil; they disobey their parents; 31 they have no understanding, no fidelity, no love, no mercy. 32 Although they know God's righteous decree that those who do such things deserve death, they not only continue to do these very things but also approve of those who practice them.

Abortions

Exodus 20, one of the "commandments" thou shall not kill.

1 Samuel 2:6, "The Lord brings death and makes alive; he brings down to the grave and raises up.

Of course, in the book of Daniel, chapter 12, Daniel prophesizes on what was to happen at the end of times which we are already seeing and experimenting.

Daniel 12, "At that time Michael, the great prince who protects your people, will arise. There will be a time of distress such as has not happened from the beginning of nations until then. But at that time your people—everyone whose name is found written in the book—will be delivered. 2 Multitudes who sleep in the dust of the earth will awake: some to everlasting life, others to shame and everlasting contempt. 3 Those who are wise[a] will shine like the brightness of the heavens, and those who lead many to righteousness, like the stars for ever and ever. 4 But you, Daniel, roll up and seal the words of the scroll until the time of the end. Many will go here and there to increase knowledge."

5 Then I, Daniel, looked, and there before me stood two others, one on this bank of the river and one on the opposite bank. 6 One of them said to the man clothed in linen, who was above the waters of the river, "How long will it be before these astonishing things are fulfilled?"

7 The man clothed in linen, who was above the waters of the river, lifted his right hand and his left hand toward heaven, and I heard him swear by him who lives forever, saying, "It will be for a time, times and half a time.[b] When the power of the holy people has been finally broken, all these things will be completed."

8 I heard, but I did not understand. So, I asked, "My lord, what will the outcome of all this be?"

9 He replied, "Go your way, Daniel, because the words are rolled up and sealed until the time of the end. 10 Many will be purified, made spotless and refined, but the wicked will continue to be wicked. None of the wicked will understand, but those who are wise will understand.

11 "From the time that the daily sacrifice is abolished and the abomination that causes desolation is set up, there will be 1,290 days. 12 Blessed is the one who waits for and reaches the end of the 1,335 days.

13 "As for you, go your way till the end. You will rest, and then at the end of the days you will rise to receive your allotted inheritance."

The bible is clear that planet earth will continue to exist after it is transformed by God and makes it clear that our Lord and the people saved, will reign here in this planet forever. Important for those believers of the climate change that think they can interfere with God's plans and are misinforming that life and the planet are coming to an end. It would be good for them to take note on what the bible says and that carries the truth. I know there are many politicians and citizens that ignore what the bible says, and many of them don't even believe, but the bible is very clear on who has control of the world.

My information for all of those that think they are higher than God the Creator is the following as the order they will occur.

2 Corinthians 5:10, For we must all appear before the judgment seat of Christ, so that each of us may receive what is due us for the things done while in the body, whether good or bad.

This is important for my catholic brothers and sisters that have been taught there is a purgatory that will rescue them on the day of the great Judgement of the White Throne. There is a young catholic preacher that says: In the bible the word purgatory does not appear, but the bible says that to present ourselves before God, we must be pure therefore, we need to purify them through the rosaries. The word trinity also doesn't exist in the bible, but Christians believe in the trinity because 1st of John 5-7 says *the Father, the Son, and the Holy Spirit, these three are one*. My information to this preacher is the following: Regarding 1st of John, it is more than clear that he is speaking about the trinity. As for the purgatory the bible says to enter heaven, we must first be pure as he states we should be. God established in the word the requirements to be transformed and to be able to be in his glory with him. I repeat:

> *2 Corinthians 5:10, For we must all appear before the judgment seat of Christ, so that each of us may receive what is due us for the things done while in the body, whether good or bad.*

This makes it very clear that no prayer can purify.

> *Hebrews 9:27, Just as people are destined to die once, and after that to face judgment,*

The following will be the judgment for all humanity:

> *Revelations 20:11-15, Then I saw a great white throne and him who was seated on it. The earth and the heavens fled from his presence, and there was no place for them. 12 And I saw the dead, great and small, standing before the throne, and books were opened. Another book was opened, which is the book of life. The dead were judged according to what they had done as recorded in the books. 13 The sea gave up the dead that were in it, and death and Hades gave up the dead that were in them, and each person was judged according to what they had done. 14 Then death and Hades were thrown into the lake of fire. The lake of fire is the second death. 15 Anyone whose name was not found written in the book of life was thrown into the lake of fire.*

To all those arrogant people who think they are beyond God should read the following verses:

Matthew 13:41-42, *The Son of Man will send out his angels, and they will weed out of his kingdom everything that causes sin and all who do evil. 42 They will throw them into the blazing furnace, where there will be weeping and gnashing of teeth.*

What is the truth about climate change? The truth is that planet earth since its creation has always had a change in temperatures, both high and low. As well there are many factors that come to play in climate changes including pollution. What is the lie or the misinformation? 1) That pollution is the only cause for changes in temperatures. 2) That controlling the pollution in USA at the cost of trillions of dollars they are going to save the planet, temperatures will go down, and hurricanes and fires will reduce. 3) That parts of the earth will disappear such as the state of Florida.

As I have already informed, I believe every person, all companies in the world, as well as the governments, are already aware that there are countries that are dumping trash in the oceans. Canada was accused by the Philippines of filling their shores with tons of trash. They must all share the responsibility of maintaining the planet as much contamination free as possible. I believe in laws and regulations that control the abuse, but I truly believe no one has the authority or power to change what belongs to God. A previously stated, God has already established what will occur with the people and the planet we live in.

ILLEGALS

The subject of the illegals is a subject that cannot be ignored even though it may sound like it is inhuman, racist, discriminating and it has been causing a lot of controversy, not only amongst politicians, but in the citizens as well. There is still much reality to analyze. There are some influential people, including legislators that are only focusing on one side of the problem, ignoring completely the other. I am clear that it has nothing to do with discrimination or racism, it is rather what the Democrats promote with the help of the liberal press and use to manipulate people against the Republicans to gain or remain in power. In this information I will give details of truths that I believe should be analyzed with all honesty and transparency, considering that many politicians, citizens and specially press, have converted the subject to racism and discrimination.

I begin by saying that there is a big difference between immigrants that have been arriving in the last decades to this country, versus the immigrants that were established in this nation 200 years ago. The immigrants that came many years ago wished to get established and make this nation theirs, bring their families, buy properties, and stimulate the economy, leaving behind their own countries. On the other hand, many of the people that immigrate today to this country come to work and send billions of dollars back to their countries, this according to what the press has shared and information that the government has shared. Of course, it would be unfair to generalize because there are still many immigrants that come to this country to make this their home. Experts have commented that the money they send strengthen other countries since it is typically used to buy and sell and even in investments. It doesn't take an expert to understand that if these billions of dollars are invested in the economy of this country instead of being sent out, the story would be different. In addition, we must consider the millions of dollars of the taxes that go unpaid for since many of them work off the books. If we add the billions of dollars they represent in the public education system, and the hospitals, we are talking about an elevated sum of money to which I will speak out a little bit further along. The following information is basically self-explanatory.

A couple years ago, the liberal press shared a video of President Trump separating the families in the boarder without any mercy. Different pictures were also shared that were captured in detention camps during the Obama administration blaming Trump of abuse against human rights. The truth is that those pictures were taken in 2014 during the Obama presidency. Was Obama wrong to separate the families? Not according to the law. The law requires that any person to cross over to the United States illegally, be detained and sent before an immigration judge where their case will be evaluated. This will determine if they will be deported back to their countries as unauthorized immigrants. If they bring their children, when the parents are processed, the children cannot go to jail with them. It is the same when US Citizens are arrested for breaking the law, their children cannot accompany them to jail. The Sherriff's Detention Center does not have a daycare. What happens is the following: after the adults are detained by the National Security Department, all minors are attended by the Health and Human Services Department and their custody is transferred to HHS within 72 hours. They are well taken care of. I might even point out that they receive better care than many of the local US children; all of this at the expense of the taxpayers. There is no Trump law that states that all families that enter illegally must be separated. What the law establishes is that all adults that cross over illegally and are captured are to be criminally processed and if it occurs to a parent, separation is inevitable. When the Obama administration tries to tend to the "family crisis" and the unaccompanied minors crossing over illegally the boarder in 2014, they placed hundreds of families in immigration detention centers. This was a practice that had stopped many years ago. The federal court prevented the families to be retained for months without a real justification to keep them there. So, Obama took the capture and liberation policy, and left many of the families free, which were released while their cases were being evaluated. Many of them disappeared in the United States instead of showing up for their cases.

Separating the families in the boarder is a result of a bad legislation approved by the democrats. To finish with the hundreds of thousands that were being captured and liberated by President Trump, they changed the policy and established cero tolerance. The policy has changed for the families that qualify and are allowed to remain together in the facilities. It must be clear that if they bring over children, and do not qualify, they will still be separated. For example, they may qualify if they have never been deported and if they are requesting political asylum. All those that do not qualify go to jail and cannot have their children with them.

President Trump was all over the media for asking about the costs related to illegal immigration. The general press revised every word and said that 250 billion dollars was an exaggeration. Nevertheless, when his argument was closely examined it became clear that he was very close to the number. The costs of illegal migration are integral. Even after subtracting the 19,000 billions of dollars in taxes paid by the 12.5 million illegal immigrants that live in the country, the cost is still 116,000 billion dollars annually to the American taxpayers. Close to two thirds of this quantity is absorbed by the local and state taxpayers who are usually the least capable of sharing the costs. One of the main boosters of the increasing costs are the 4.2 million immigrant children that automatically become American citizens. As a fact, to the US Citizens it costs $45 billion in education, both local and state annual, without mentioning the billions of dollars in welfare. We should also point out the 30,000 billion dollars in medical funds that the families that are not citizens have better probabilities to receive relief payments than the families that are US Citizens. One half of the non-citizens receive Medicaid in comparison with the 23 percent of the native citizens, while half of the non-citizens are on food stamps. What worries the most is that the non-citizens that stay for long term are more propense to make use of these programs versus those that just arrived. More than half receive welfare and it increases to 70 percent for those that stay longer than 10 years. The additional load of illegal citizens is $600 on each US citizen annually. The institutional load of illegal migration also includes a criminal statistic four times greater.

Of all federal prisoners, 26 percent are non-citizens of which two thirds are in the United States illegally. Considering that it costs the federal government $32,000 annually for each prisoner, the approximately 25,000 non-citizens in our system represent close to one billion dollars in annual costs, without mentioning the costs of the prisons and the application of the immigration laws. The general surveillance figures have also gone up in the boarders. The number of agents in the border patrol has increased five times in the last 25 years and has almost doubled in the last 15 years. In the meantime, the costs of protecting the border with Mexico has multiplied almost by ten in the same period of 25 years and almost $4,000 billion annually. This without taking into consideration the 43 percent of the illegal immigrants that do not present themselves to their court hearings after their detentions.

We also must inform that President Trump was accused of being evil by being wrongly accused by DACA of costing money and causing problems to the US Citizens. According to the misinformation that is being shared, the DACA in majority, were studying in the university and were not any trouble. Unfortunately, the reality of the HLS is another. The illegal immigrants represent more than seven thousand prisoners for battery and assault, 875 for sexual assault, 8 for murder and more than 7,000 for drugs and driving under the influence. For these and many other reasons is why the president has opposed to the illegal migration, it had nothing to do with racism. The racism aspect was mere speculations and political interests. I am not saying he was a saint, but I have for certainty that the liberal press and the democrats have created a wrongful image with the intention of taking him away from power.

With all honesty, I wish they could all become legal citizens and live the way others live. But we must have the responsibility to analyze all the positive and negative aspects and place them in balance to establish who is right. This is the natural aspect because there is another that no one seems to consider and it's the spiritual and legal part. Things are not as easy as many people or Christians think, that is why I mentioned at the beginning what the Word of God teaches. *"Whoever wants to be my disciple must deny themselves and take up their cross and follow me. The liars and disobedient will not inherit the kingdom of God."* Those that are here illegally are here is disobedience and based on lies.

Romans 13:1-7, Let everyone be subject to the governing authorities, for there is no authority except that which God has established. The authorities that exist have been established by God. 2 Consequently, whoever rebels against the authority is rebelling against what God has instituted, and those who do so will bring judgment on themselves. 3 For rulers hold no terror for those who do right, but for those who do wrong. Do you want to be free from fear of the one in authority? Then do what is right and you will be commended. 4 For the one in authority is God's servant for your good. But if you do wrong, be afraid, for rulers do not bear the sword for no reason. They are God's servants, agents of wrath to bring punishment on the wrongdoer. 5 Therefore, it is necessary to submit to the authorities, not only because of possible punishment but also as a matter of conscience.

6 This is also why you pay taxes, for the authorities are God's servants, who give their full time to governing. 7 Give to everyone what you owe them: If you owe taxes, pay taxes; if revenue, then revenue; if respect, then respect; if honor, then honor.

> *1 Peter 2:13-14, Submit yourselves for the Lord's sake to every human authority: whether to the emperor, as the supreme authority, 14 or to governors, who are sent by him to punish those who do wrong and to commend those who do right.*
>
> *Titus 3:1, Remind the people to be subject to rulers and authorities, to be obedient, to be ready to do whatever is good,*

Proverbs 12:22, The Lord detests lying lips, but he delights in people who are trustworthy.

Revelations 21:8, But the cowardly, the unbelieving, the vile, the murderers, the sexually immoral, those who practice magic arts, the idolaters, and all liars—they will be consigned to the fiery lake of burning sulfur. This is the second death."

There is a big number of reporters congressmen, and other people that have influence in the country that are only looking at one side of this problem. Let's look closely at the following example and analyze it with transparency and honesty. Any person that has have to go to the immigration building knows that they must get there early, and even more it would seem they have to be there the day before to get assistance. Now, imagine that I stayed home sleeping and when I wake up, I get there and go to the head of the line, and I get attended first without any consideration for the people that were there before. It is also important to inform that there are millions of people that have traveled a long distance from their country of origin towards the embassy and have paid large sums of money and have been waiting for years to receive a legal entrance to the country. It now results that those that have cut ahead in line illegally have to be attended with priority, and those that have been waiting for years and have invested money in the process, have done things legally and by the book have either been left behind or cut off completely.

I am sure that these reporters and congressmen and people of influence in television, for many reasons, are completely in the wrong. They are only focusing on the humanitarian aspect and are not taking into consideration the legal aspect, which is equally important. For example, if you are in another country and want to come into this country: why request legal entry, follow the laws, pay money, and go through the hassle of the wait? It's better to enter illegally, seemingly easier, faster, and secure. I have also heard the excuse that this country is the country of immigrants. It is a fact that this country, like many others, is composed of immigrants. If we truly believe the biblical information, we believe that since the beginning of humanity, or after the flood, when Noah and his children filled the earth again, we know we are all descendants of the same family. In Genesis 9:18-29 we find the story of what occurred after Noah and his family stepped out of the ark.

The sons of Noah who came out of the ark were Shem, Ham and Japheth. (Ham was the father of Canaan.) 19 These were the three sons of Noah, and from them came the people who were scattered over the whole earth.

20 Noah, a man of the soil, proceeded[a] to plant a vineyard. 21 When he drank some of its wine, he became drunk and lay uncovered inside his tent. 22 Ham, the father of Canaan, saw his father naked and told his two brothers outside. 23 But Shem and Japheth took a garment and laid it across their shoulders; then they walked in backward and covered their father's naked body. Their faces were turned the other way so that they would not see their father naked.

24 When Noah awoke from his wine and found out what his youngest son had done to him, 25 he said,

"Cursed be Canaan!
  The lowest of slaves
  will he be to his brothers."

26 He also said,

> *"Praise be to the Lord, the God of Shem!*
> *May Canaan be the slave of Shem.*
> *27 May God extend Japheth's[b] territory;*
> *may Japheth live in the tents of Shem,*
> *and may Canaan be the slave of Japheth."*
>
> *28 After the flood Noah lived 350 years. 29 Noah lived a*
> *total of 950 years, and then he died.*

As I have previously mentioned in the bible story, in the book of Genesis we see that Shem stayed in the Middle East, Japhet migrated to Europe, and Ham to the south. We know that from Africa, Europe and the Middle East, people have immigrated throughout the world, and we have linked ourselves to the three offspring, except for a group of Jews that have not been allowed to mix themselves to any other descendants.

Why do I mention that the saying of this being a country of immigrants is out of context? Because it is a country of immigrants, but of legal emigrants and not illegals. In the context that the liberal press and the democrats have tried to communicate is that since it is a country composed of immigrants, they come here, and they must be accepted and given the residency. If we are honest, no other country in the world accepts this. That is why there have been laws created in other countries so they could guide us. All countries have migration laws in place to control their borders and their population. In the case of our country, most of the immigrants are Mexicans, and Mexico is one of the countries with very strict migration laws. The press has reported how Mexican government officials have deported Cuban immigrants without any mercy. There has also been mention of many people from South and Center America that have died in the hands of Mexicans trying to cross the border. How many women have been raped? There has been a lot of abuse to people

71

that have crossed their borders, including kids. I asked Jorge Ramos and Maria E. Salinas since they are Mexican, why they do not talk about this instead of attacking the republicans of this country so enthusiastically as they do? It's incredible the hypocrisy. What is surprising to me is to see democrat congressmen trying to reward those that violate the laws that the same government has established and accusing the republicans of being racist and against immigrants. All this because they reinforce the laws established precisely for our own well-being and by the same congress.

The following is an example of many

During the President Clinton administration, democrat by the way, signed the following law: Immigration Reform Act of 1996. This law clearly states that if an immigrant is here in the country residing illegally, they would not be eligible for any financial help or support for education purposes (tuition). According to the press releases that have been shared, there are 11 states in clear violation of this federal law that they deciding to change it at state level. The federal government, for political convenience, has done nothing about it. We all know those political conveniences. Anything that jeopardizes immigrants represents less votes during elections of their family members and from those that supports them. What are these 11 states doing? They are creating a special privilege for non-citizens versus legal residents in the country. There are millions of young people in the rest of the states that do not receive secondary financial support for college and as a result, many of these people can't afford to attend college. Meanwhile, many illegal residents are enjoying these benefits and many US Citizens can't. This is the case of one of my daughters that for her to assist the university we have had to take loans of thousands of dollars. The illegal residents and their supporters take part in manifestations and protests screaming and asking for justice. I ask myself the following question: What justice are they talking about? Cases like the one I just mentioned are the ones that precisely have the legislators and many us citizens irritated, this mainly since they feel offended by the illegals and their defenders. They are here breaking the laws, and at the same time, claiming rights they don't have. It has come to that point that they already have an organization that is being coordinated by a young woman who was interviewed by the Fox News Presenter Bill O'Reily, where she was demanding that the illegal immigrants not be called illegal. It is incredible and unacceptable to listen to the activists of this organization. The refuse to be called illegal immigrants even though they are here illegally.

There is also the humanitarian aspect that must be given importance and be careful in how the laws are applied that can harm innocent children, who are not at fault of the bad decisions their parents and adults make. I mention this because all people that enter the country illegally are perfectly aware they are violating the laws, and know they are at risk of being captured and deported or even incarcerated. They should be honest and stop blaming the legislators or the Americans about their bad decisions. I also think that the activists and reporters that defend the immigrants' rights should be more intelligent when treating this matter, the US Congress is the only one with the power to resolve these issues. Attacking one political party or the other is not the correct matter to gain support. I believe they should rest on the fact that the democrats will solve this immigration situation for the following reasons.

President Obama and the democrats promised to do an Immigration Reform if Obama was elected. In the year 2007, the Democratic Party gained majority in both cameras and in the last two years up to 2010, since President Obama was elected. What did they do?

Absolutely nothing. When elections come back, I can assure you that the same type of promises will return. That's the way it has always been, but their defenders have always acted blind so the situation. Nevertheless, the president that did do something for them was a republican, President Reagan. The other president that also tried to solve this situation was another republican, George W Bush. He couldn't do anything because the Democrats that promised the reform opposed. These are the realities that the liberal press and the democrats do not inform, instead they continue to fool a big number of people that refuse to gain true knowledge. What I share here are simple truths that the liberal press will not inform nor will the democratic party. I hope that this problem can be resolved favorably for the greater good. Something else that the legislators need to know is that if this problem is not fixed and they deport all of whom are working in agriculture, believe it or not, tomatoes will cost us $10.00 the pound as well as the lettuce. I am aware that no legal resident or US Citizen will do that labor for $7.50 the hour and without benefits. We know they are extremely hard labors.

THE BELIEF THAT THE DEMOCRATS IS THE POLITICAL PARTY FOR THE POOR, AND THE REPUBLICANS FOR THE RICH

There are realities that should be analyzed with all honesty and transparency. It worries me when I see politicians during their political campaigns offer social benefits at the cost of many trillions of dollars like the case of climate change, health insurance, and many other benefits offered, and see multitudes of citizens applauding those false promises and blindly believing these impossible promises. The following information stresses the importance to be well informed. Since I was a kid, I have heard that the Democratic Party is the one that represents the poor, and the Republicans represent the rich. Personal experiences have taught me along the years the contraire.

I was raised in a humble democratic home. I remember when I voted for the first time, I voted under the traditional views of my parents and family members, as well as the belief that I was voting for the people that would help me. I want to make a point that in those times, the democratic party was functioning and had a different ideology; but it has been changing at gigantic, scandalous steps taking an opposite moral road and everything that has been established by God. Something that was taught to me was the democratic party was for the poor and the republicans for the rich. After many years I started to see that what was taught to me didn't seem to add up to what I was seeing. I started to analyze everything with an open mind, honestly and with transparency. In many analyses that I have done based on personal experience and based on reality, I realized those lessons are currently incorrect and they lack truth. I have no doubt that those teachings and beliefs are still being imposed due to the lack of information and knowledge since several worried citizens continue anchored in the political traditions and convictions. I must make clear that this information I did not obtain it from the press, but rather from my personal experiences, from viewing for the past 22 years the bad behavior, conning the social programs, many years of studying, government sources, and history books.

The information that the Democratic Party if for the poor is not my opinion due to the following reasons:

I want to use my own example since it's the reality of all the middle class. The following experience talks about itself. It also speaks on the rest of the information that will be shared and will make the truth clearer.

Living in New York in 1988, in a fire my family and I lost everything. I only had my work uniform and my family the clothes they had on at the moment. The fire destroyed everything, and we were left stranded on the street. I was advised by my boss, family members and friends, to visit social services and request help. I went to several government agencies and all help was denied because of the salary I earned. My salary as a mailman was $33k at that time. This was the same reason that was used to deny my daughters to participate from the Free Lunch Program at school. The reality is that the middle class is only good for paying taxes. The following explains the reasons I have for calling many people slick. I worked for 22 years as a mailman in New York and Florida. I will never forget that in a small area, I used to deliver hundreds of social helps checks representing billions of dollars monthly. What was interesting is that in majority, they were young women with their husbands living at home, lying to the government by saying they lived alone. Just look at the way the middle class pays taxes today, and these con artists live off housing programs, food stamps, SSI, and many other government programs that many of you know. This in addition to the many people collecting money from being falsely disabled, many for back problems or mental conditions. In all honesty, I know very well what I am informing here. What is completely clear to me is that I am not rich, I worked for 24 years for a private company, and 22 years at the US Postal Service. I have worked since 1962 to this day. I am thankful that I never had to step into an unemployment line. Aside from the money that President GW Bush handed out to stimulate the economy, no other Democrat or Republican President had given any help, and I have not qualified for anything else. President John F. Kennedy, one of the most outstanding presidents of the Nation, despite representing the Democratic party, whom 98% are liberals, he was one few conservative that I am referring to (and one of my favorites) stated the following: "it's not what the government can do for you; it's what you can do for the government." Unfortunately, many people today want to do the opposite to what Kennedy said, including President Obama and most of the democrats. If we are honest, these are completely honest facts. They want food stamps, housing, health care, cellular phones, without having to work or do anything. The way the Democrats have the country, soon based on the social justice they have anchored on, they will buy them cars so they can move around. The following is what God has established.

> *2 Thessalonians 3: 6-12, 6 In the name of the Lord Jesus Christ, we command you, brothers and sisters, to keep away from every believer who is idle and disruptive and does not live according to the teaching[a] you received from us. 7 For you yourselves know how you ought to follow our example. We were not idle when we were with you, 8 nor did we eat anyone's food without paying for it. On the contrary, we worked night and day, laboring and toiling so that we would not be a burden to any of you. 9 We did this, not because we do not have the right to such help, but in order to offer ourselves as a model for you to imitate. 10 For even when we were with you, we gave you this rule: "The one who is unwilling to work shall not eat."*

> *11 We hear that some among you are idle and disruptive. They are not busy; they are busybodies. 12 Such people we command and urge in the Lord Jesus Christ to settle down and earn the food they eat.*

Regarding the saying that the Republicans are the party for the rich and the Democrats for the poor, it is false. When we analyze the fact that no one that wins $30,000 a year or more, qualifies for any government help. When I say ANY is ANY, cero. For example, my wife suffered a brain stroke leaving her permanently disabled. She lost mobility to her whole right side. She also suffers another condition for which she will need to take medications for life. I am retired and no one believes me when I say that we do not qualify for any help for her. Not Medicare, nor Medicaid, absolutely nothing. Not even for the medicine that costs us in monthly copayments of $200.00. All petitions I have requested have been denied on paper because of my financial earnings, and we as the rest of the middle-class citizens, are poor. With great sacrifices we survive every month. How is it then that the Democratic Party is for the poor? Without a doubt, if I had done what many people do to receive the help, we would have received it a long time ago. For example, a person that I know that fools the system, told me the following: If you want the hep for her, since your earnings don't qualify you, get a divorce. The divorce will divide your earnings by half. That half will belong to your wife and with the new earnings, she will not be denied. Of course, as a Christian man I would never do anything like that. It is better to enter the heavens poor and with need, than with all that money to hell. Besides that, I have always been a good citizen. I mentioned the example just to show you how many people outsmart the system. What I can recognize easily after 22 years of service in New York and Florida, going house to house, is a big number of people like the person that advised me to evade the law, that live thanks to the help they receive through lies and deceit. This is just one of the many ways people do to receive help from the government. The problem is what the Democratic party established in the beginning was correct, but it has then been corrupted, and they have turned it into a lifestyle. The purpose was to help the truly needy, who I also agree should receive the help. Many of the people that lie to get these funds are also poor, but they do not want to strive to work as hard as the rest of the poor. They should sacrifice themselves like many do with several jobs, and sacrifice other things to push forward, without having to depend on the taxpayers. The problem today with the democratic platform is that they take away from those that have, to give to those that don't. We know of millions, not in real need, but who outsmart the system, that are currently living at the expense of the taxpayers. I don't want to

sound hateful; I know that there are people who truly need the help and biblically we are called to do so. But this should only be for people who really needed the help and not to millions of people I saw every day in various communities in New York and Florida, who sat under the trees drinking alcohol and smoking marihuana, with the money from the taxes I paid and the working class. This is what the husbands of the women I previously mentioned, did. This occurs in the whole nation. That is why I still can't comprehend how the middle working class, votes for the democrats that support this behavior. What I believe is that they lack knowledge, their political convictions blinded them, or they are medicated. Are they in agreement to what God has established? The answer is no. What God established is the following: *"By the sweat of your brow will you have food to eat until you return to the ground from which you were made."* Not only did God establish this, but President Bill Clinton also had problems with the democrats that accused him of betraying poor children with the following reform.

In 1992, the then elected president, promised, and said they needed to do a reform to the welfare program, because in his own words, there were millions of young women living of this help, and this was costing the federal government billions of dollars. These were the checks I mentioned I delivered when I was a mailman. The cost is paid clearly by those that work and pay taxes. On August 22, 1993, after working on this reform with the Congress Leader, Republican, Speaker of the House, Newt Gingrich, signed the reform that states the following:

1) Every person that receives welfare had a maximum of 5 years so that the assistance could terminate.

2) Every person that received a check from welfare and was capable to work, had to find a job within a term of two years. It was approved to give the states incentives so they could help the people that where in need, and the sole responsibility fell on the state. That is how they put an end to the billion-dollar abuse of funds every year. That is how millions of jobs were created as well since the millions of people living of the taxpayer's money had to seek out jobs. In reference to this reform, President Clinton said: "today we are taking a historic change to make welfare what it was meant to be, a second chance and not a way of life."

THE BELIEF THAT THE DEMOCRATS IS THE POLITICAL PARTY FOR THE POOR, AND THE REPUBLICANS FOR THE RICH

THE TAXES

This is the total opposite of what was promised by the elected president Barak Obama of taking from those that have a lot, to give to those that don't. Fulfilling his promise, he increased the contributions to those that earned $250k or more a year. The reality is that on paper this sounds good, but the facts are that people who earn $250 thousand dollars or more a year, only represent 5% of the population. Even though this 5% make a good contribution, it's with the contribution that comes out of pocket of the middle class that most government programs are sustained. It is so that an expert in economy informed that the federal government collects almost $800 billion in taxes from the citizens, and $200 billion from corporations.

We must also analyze correctly the information shared by the press and political media; the following is an example of many. President Obama and the Democrats, as well as the liberal press, say the rich pay the same amount of taxes as a secretary. This information is incorrect. There are cases that the rich pay the same percentage as a secretary, but not by any means, does it indicate they pay the same amount. As an example: If we use as base that both pay 15% of taxes, and the secretary earns $50,000, 15% represents $7,500. A rich person that earns $1,000,000, with the same percentage of 15%, his payment would be $150,000. If it is two million the quantity would be $300,000. The higher the earning, the higher the contribution. The information that they don't pat the Fare Share, in my opinion is incorrect and merely political. We also must mention that 49% do not pay taxes. Besides that, the vast majority of those that pay taxes by credits and expenses, as well as other costs, we receive back a good portion of what we pay.

The following information is based on a comment that President Obama made. The churches based on Malachi 3:10 request 10% of their profits to all their members. My question based on the word of God, is the following: Can the church request the tithe based on salary? Can the church request 10% to those that earn $30,000 and those that earn $100,000 a 20%? In the book of Hebrews, chapter 7, Abraham – the richest man of those times, paid his tithe (the 10%) to the king and priest Melchizedek, as did the rest of the people that paid their tithe to the priests, as per the law. This was what God established since the beginning. In church, to pay the 10% despite their earnings is correct. This is also true in the government taxes, those that pay the stipulated percent is also correct. That is honestly the Fare Share. Believe me, it does not affect me or benefit me, what rich people pay, quite the opposite, it affects me more what they deduct on my earnings to give to the slick people, because part of my taxes are directed towards these funds. I mention this because it was Obama who took a Bible verse out of context. He said the Bible mentioned that he who has more, more will be asked of him. That is what the Bible teaches but not in reference to money, but rather work.

I don't think anyone likes for the government to increase their taxes, especially if we earn little money. Nevertheless, when a candidate from our party states that if elected, they will increase our taxes, we don't care and we give him our vote anyways. The reason is very simple, If we are democrats, at no expense will we give a republican our vote. If we are republicans, at no expense will we give a democrat our vote. The roots of our traditions are extremely deep, so deep that we do not sit down to analyze, think, or listen. And this is not just on taxes, it's in general terms. I have sustained conversations with citizens, including democrats, that only know how to listen through the liberal press, since they are the only media, they listen to. They have been brainwashed since they are not laborers; they are activists of political parties.

In the presidential race between President Barak Obama and Senator John McCain, it was a clear example. President Barak Obama was clear that he would increase the taxes. Of course, he was clear when he stated he would do it with the people that earn $250,000 or more a year. I say people because even though he spoke about small business owners, those that earn $250,000 through that business, had to pay the taxes. I have no doubt that if President Obama told everyone that works that he would increase their taxes, the results would have been the same.

Regarding the taxes, the difference between democrats and republicans is the following. The republicans, or the conservatives, will not legislate for any reason a tax increase. Even though they are accused to defend this just for the rich, they do not allow a tax increase for anyone. The reasons are simple, the republicans, conservative believe that only the minimum necessary should be paid to cover the government needs, so that the government can assume their responsibility, no matter what the cost, if its justified. The democrats on the other hand, the majority, like to increase taxes for every existing program. This is the reason they are identified as the big, liberal, government. When I talk about the position of both parties, I am referring to federal, state, and local aspects. Why do I say that the Democratic Party is not for the poor but for the slick? You must live in big cities like New York, Los Angeles, Chicago, to notice the quantity of people that live off fabricated, democratic programs. This is the reason why the Republican party doesn't have an opportunity to win elections in these states. In these states where the "so called poor people" mainly live, that I am aware there are truly poor cases, the truth is there are more slick people living off the working class, that work hard and costs billions in taxes. If you want to be better informed, look closely at the housing program, better known section 8, or the food stamps program. Despite existing people that truly need it, the number of those that take advantage of the help with lies and deceit, is bigger.

In all honesty, let's analyze the following:

People are always complaining about the constant cost increases in gasoline, food, insurance, housing, and taxes. What doesn't increase is the salary. If we analyze without the politics, the truth is that when they increase the taxes to businesses, it is guaranteed that they will pass the cost to the consumers. Therefore, when in President Obama's campaign, I heard the arguments of many people stating that Obama would take away from the rich and give to the poor, it just made me laugh for the following reasons. Like I mentioned, any business that receives a cost increase will pass this on to the consumer. If it's a manufacturer, they will pass this on to the products or it results in layoffs and employee reduction. It is also true that there are many people that earn millions of dollars in sports and other jobs that can't pass this on to anyone. But these people amongst the 330 million citizens is nothing. These are realities that people don't analyze because they are blind by their beliefs and traditions. They only receive information from the liberal press or information from the internet, which should also be questioned if credible or not, and if it's reliable. What is guaranteed is that the consequences are paid by the middle-class, and those that work hard, sometimes with 2-3 jobs, just to push forward.

The reason why I say the middle class is because, the truly poor receive some type of benefits. But the working class are the ones that will end up paying these costs directly or indirectly. For example, were you aware that we pay 46 different types of taxes? Did you know that on the telephone bill itself we pay 7 types of taxes? Some are called fees and other taxes. Did you know that one of these taxes is by the federal government so that we can pay the service for those that are poor and receive this type of help? Did you know that if you have 2 telephones, one at your house and the other a cellphone, you are paying twice for this purpose? Correctly stated, everyone that has a cellphone is paying the service of the previously mentioned. You should also know that the annual cost of these phones are more than a thousand million dollars.

Its unexplainable the number of citizens that do not analyze these types of facts, and they boast saying that Obama will take away from the rich to give to the poor. Oddly enough they are the first ones to complain when there are all types of increases, in the supermarket, or at the gas station, housing, in everything. This is as real as 2+2=4. That is why the rich will continue to be rich, no matter what gets taken away; and the poor continue to be poor. For example, the Obama administration imposed more taxes to the gasoline corporations; To whom do these corporations pass this cost to? Of course, to the consumer. This is one of the reasons that prices go up. When I say all prices go up, it's all prices, because we all know that the gasoline industry moves everything. Another reason is that the same government raises the prices on many things since the consumer pays taxes for the use of gasoline. Up to February 2011, we were nationally paying 48.1 cents on taxes on gas per gallon and 53.1 on diesel. On top of this tax, the state and different communities charge tax for gasoline use. Like I already explained, when they charge the gasoline owners a higher tax, they pass this charge onto the consumers. They can't increase the tax on the gasoline, but they increase the cost at the pump. In a housing building when the taxes are increased, or the insurance, or the repairs… who do you think the property owners will pass the cost to? Isn't to tenants when they pay rent? What I have clear is that no matter what the business is, it will not cost them a cent because all cost increases will pass onto the consumers. That is why I do not agree with any type of increase because it will end up coming out of my pocket. The average consumer only thinks about the taxes that are deducted directly from their check and do not analyze these other factors that are drowning us, and each day that continues to pass by with these types of government, will get a lot worse. If you really want to know why not so long ago a gallon of milk cost $1.99 and an apartment went for $400 a month, and now the gallon of milk goes for $4.99 and the same apartments for $2,000. Another question we should analyze is: What is the main reason that millionaires and billionaires donate millions of dollars and support to Democratic politics? Aren't the Democrats the ones that want to increase greatly the taxes? I have already given you the answer.

THE ABUSE AND POWER OF MONEY

In this country money is wasted by trillions and trillions at the expense of who pay taxes. As a good citizen I am aware of the taxes that need to be paid, but the problem is on how the government uses the money; not just on what already been mentioned, but in general. The list is big of the things we see daily everywhere you go, or at a government office when you need to resolve something. What I am mentioning is at Federal, State, County, and the City.

Politicians don't care about the citizens, they continue to approve tax increases at any cost, to obtain money for all their unnecessary expenses. I live in Palm Beach County in Florida. As many know, the value of properties came down a lot of money, as well it is known we pay property taxes. The state increased the discount on property tax (Homestead Exemptions) from $25 to $50 thousand dollars. With the decrease of the properties and the discount that the state gave, my taxes increased $500.00 more of what I had to pay per years instead of going down for all the reasons I just explained. What is the excuse the legislators give? That all the departments in the county increased their budgets. Like I mentioned, I live in Palm Beach, and I see daily cars from other counties; police cars, and other city cars that live here in this county and use these cars for personal use, at the expense of the citizens. Believe me, it would take over 100 books to write about the way these governmental departments waste money wrongfully. This happens in all the nation. Just to mention an example, if you pass by an area where a hole is being made to place a pole, you will see 6-7 workers. One is the engineer, the other one opens the hole, the other one takes out the dirt, etc., etc., etc. The same happens in construction and in road repairs. If we had responsible governments with the trillions of dollars that circulate, and that are obtained annually through so many taxes. We even pay for the air we breathe. The country wouldn't have a debt of so many trillions.

Many legislators and a big number of citizens are not paying attention to the country's debt, but the higher the debt, the less value the American dollar has, and the less credit line it will have to borrow. Therefore, for the first time the country's credit went down from AAA to AA. They are also threatening to decrease it to A. All the empires of the world have fallen and regardless of how confident we feel, this will be no different. The other 5 empires thought the same.

Another example of how money is wasted is the following. Of the $800 billion dollars that the Obama administration gave out, trying to stimulate the economy, they were on the news of cities that fabricated sidewalks in places that people didn't even walk, since they were impassable. The excuse they gave on the press was that it was better to give people a job that needed one, than having to give back the money from not using it. The reality is that there are people that are ok with this abuse, but I am not. It is completely unjust that we work so hard to get deductions on paychecks to support this type of injustice. I am sure that even though people might not have all the information, they are aware of all the money that is wasted. I am also sure that even when the legislators of our party run a trailer over us, we will continue to support them. These are the reasons why of the abuse of power. How is it that the democratic party is the party for the poor, and they gave away the money from the $800 billion dollars to multimillionaire companies, and the poor property owners were left in limbo? Many homeowners lost their homes, not because they couldn't pay for the mortgage, but because the property taxes were so high, they couldn't afford it.

The democrats are mainly the ones that have the population and homeowners paying so many and elevated taxes. This is something that no one can deny. For decades, this has been a war between Democrats and Republicans. At this moment, the reason why the Obama Care law took so long and one of the reasons that the Republicans unanimously voted against the law was because the Democrats included the use of the money from the taxes to pay for the abortions. The government was almost forced to shut down their operations because the Democrats wanted to eliminate the tax reduction that President George W Bush had granted. The excuse they used was that this only benefited the rich. These are the excuses that people don't analyze. At the end of the day, President Obama admitted that they needed to continue with the reduction of tax exception, because if not it would affect all who worked and paid taxes, and in the situation that the country was in, it wasn't convenient. What you just read, were the very words of President Obama. Why didn't the liberal press attack President Obama for continuing the tax cuts to the rich, as they say, that President Bush did? I agree with President Obama because he did what was correct, but the reality is that because it was Obama, and they support him, there seem to be no issues and they don't accuse him of supporting the rich. Another evidence that conforms the Democrats are responsible for the increase of so many taxes, are the states that have been dominated by the Democrats like New York, California,

Illinois, and others including other cities dominated by the Democrats, the taxes are so elevated that there is no room for the middle class, and they are moving from these places by the millions because it is impossible to exist.

THE BIAS OF THE LIBERAL PRESS

In reference to the liberal press, in all honesty we must analyze the following: in 2006 the gasoline went up to $4.00 a gallon and the press were on top on President Bush and accused him of the price increase and there would be a daily debacle in the press. Now, under Obama's administration the price of gasoline that was at $1.84 in January of 2009, have surpassed the $4.00, but because it's President Obama, who they support, you don't hear anyone from the liberal press attacking Obama or blaming him for the gasoline cost increase. Like I have been saying, the liberal press is partial zed with the Democrats. They inform what is convenient to them, they don't inform, or they twist the truth. The following is an example. President Trump ran for presidency with a long list of promises, which he has fulfilled to perfection. Starting with the wall on the boarder, decreasing taxes and creating millions of jobs, reflecting on the lowest numbers of unemployment amongst the afro American and Latin community, as well as in women. He eliminated thousands of regulations that were killing employments. He brought unemployment down to 3.5, he achieved an employment record that had not been seen in 50 years. He built the economy that Obama had not been able to in 8 years. As a result, the investments in Wall Street have been breaking all records, which in turn has helped the pensions of the workers and all inversions. Through tax cuts he has been able to impact an increase in salary. He was able to eliminate the mandate of paying the fine of Obama Care through tax payments at the end of the year. He doubled the tax credit for families alleviating the load. As a result of everything that has been mentioned, the levels of poverty went down from 14.8 that Obama left behind, to 10.5 in his first three years and a half. He did several amendments, amongst them to the veterans that were suffering the medical attentions that they desperately need. He has accomplished many companies to come back to the country. There have been over 700 good accomplishments for the help of many citizens, in addition to many other accomplishments in the exterior politics, like the treaties of Mexico and Canada, and the billions of dollars that are coming in from China through the rates he has imposed. All the information stated above is correct. The question I ask to be answered honestly is the following: What liberal press has given him any credit for these achievements? The answer is never. Quite the opposite, they only criticize his achievements.

Like I informed before, the liberal press and the hypocrites in the Democratic Party, have been on top of the President for the attack of the General in Iran, who was eliminated because of his attacks to US Embassy in Iraq. This General has been directly involved in terrorist attacks against American soldiers in Afghanistan, Iraq, and other African countries. Besides having killed 600 American soldiers. He was also responsible for other deaths including attacks in Israel. Just a couple of months ago, Iran dropped a drone on United States on international waters. In the last year they have captured 6 ships. They also took hostage a group of soldiers also in international waters. Why the bias and the hypocrisy of the previously mentioned? The same thing that Trump is doing, is what the last 4 presidents have done as well. President Obama ordered a total of 563 attacks, in large part by airplanes not manned; they pointed towards Pakistan, Somalia and Yemen during his two cycles. This is comparison to 57 attacks under President Bush. Between 384 and 807 civilians died in those countries as per data reported by the government office in these countries. Obama also started an air campaign targeted towards Yemen. His first attack was a catastrophe: the Commanders thought they were pointing towards Al Qaeda, but in its place, they pointed towards a tribe with racism ammunitions, killing 55 people. Twenty-one were kids, and ten of them were under the age of five. Twelve were women of whom 5 were pregnant. Pakistan was the center of operations of drones on President Obama's first mandate. The rhythm of the attacks had accelerated in the second half of 2008 at the end of Bush's mandate. After four years the attacks were reduced to occasional attacks. But the following year when he assumed his charge, Obama ordered more drone attacks than Bush during his presidency. Of these attacks, 54 of them occurred in Pakistan on the year 2009.

I wonder why the liberal press and the Democrats are more honest and ask themselves: what is a General from Iran doing at a war camp in Iraq, were Americans are currently helping under the permission of Iraq, against the terrorist acts of ISIS? Quite opposite, they are accusing President Trump of abuse of power just by eliminating a US enemy. I can imagine that meetings behind closed doors have already started to plan the next impeachment. It is my opinion that any citizen that doesn't see the bias and the delirious truth, is not honest. It is more than clear based on the real events and the facts of both presidents, that the liberal press is bias, and is not honest, and less credible. It is totally incredible that having killed the terrorist General from Iran, he gets accused of criminal, while on the other hand, Obama doesn't get criticized for having killed all those children and women.

THE OBAMA CARE HEALTH INSURANCE

If we analyze the Obama Care health insurance, we will see that it affects more people than it helps. In my way of seeing it, it's a disaster that the Democrats have done with this famous law of the Obama Health Insurance. The liberal press will only inform what is convenient to them. In my case, and it's the case of millions of workers, my insurance premium has gone up $200.00 more a month. In addition to this, all insurance increased the copayments of medicines and doctor visits. Therefore, after a year of this law being active, more than 63% of the citizens, disapprove of it and wants it to be eliminated. The press and the Democratic legislators inform the good that this law is, because it gives them coverage of insurance to their parents and all children up to the age of 26. What they don't report is that now, all the insurance companies must enlist millions of youths in their parent's policies, and this will cost millions of dollars. That money must come from somewhere, and of course in this case, from the parents. Like I had informed, no company or no one for tax reasons or other reasons like this law, that costs more money, will for sure pass it on to the consumer. Something else that they don't report is the following: the way this law was created in reference to the coverage, it doesn't matter if the children are married or not, if they still live with their parents, wither they have a job and their employer provides health insurance, the insurance of the parents must provide coverage for their kids under this law. It sounds ridiculous, but it is so. For example, I have a 23-year-old daughter that works, and where she works, she has a health insurance. The advice that the employers gave her was that if she obtained insurance with them, she was throwing away money, since automatically my insurance would cover her. And exactly that is how it was. I obtained written information from the insurance company and from the government that passed the law, explaining in detail the new law, its coverage, responsibilities, and obligations. Like I stated, this law affected more people than it benefitted. We go back to the same thing; they do this to benefit the group of people I mentioned before. There were 33 million people that didn't have health insurance, that is the reason why this law was created, but not even with this insurance does all the population get insurance. A lot of them do not want to work and have made the government programs their way of life. They are the ones who will receive health insurance with the tax money that we pay in taxes. I inform this because in 2015, the information is that the Obama Care cost the taxpayers more than $56 billions in suicides alone. There is also a percentage of people, mainly youth, that work in places that do not provide insurance, and that in certain cases will be benefitted, and in others disadvantaged, since with the little they gain, they can't contribute to this mandate. If they don't buy health insurance, they will be fined. I say fined because, for it to sound better, they will call it taxes. The reality is that it is a fine that will be charged through taxes. The mandate of this law dictates that it is compulsory to buy health insurance for those who work. The mandate in this law is currently under process in court because many experts, including judges' rulings, find it unconstitutional. It is based on the amendment number 14 of the constitution that was proposed on June

13th, 1866, and ratified on July 9th, 1868, that says that no state can create or impose laws that take away the privileges that each citizen has, their life's liberties, and their property rights. That mandate violates this amendment. The Democrats to fool this amendment, since the law talks about fining those that do not buy insurance, and by fining them they would violate this amendment of the Constitution, they changed the term fine to a tax based on article 1, section 8, of the same Constitution that says the Congress has the power to create laws that collect taxes from citizens. The 16th amendment of the Constitution that also gives the Congress the same power, to collect taxes from any identity without the importance of the area represented that the states have or its population. Another trick they did to pass the law, was using the Budget Reconciliation process. This is supposed to be only applied for a budget legislation. The reason why they used this process that the constitution requires, in an amendment that the Senate passed in 1975, that clearly states that all laws of this magnitude that affect the citizens, must be approved by a third part of the Senate, that would be 60 votes. The reason why 60 votes are needed according to the constitutional tale, is because the Congress and Senate, are elected by the people, therefore, when they deprive the Senators their rights, they are affecting the people who were the ones to elect the Senators so they could in turn, make decisions for them. In other words, this law was imposed by the Democrats by force without considering the constitutional rights that the citizens have. How did they achieve it? They are the majority in both chambers and having the presidency. They made the mentioned changes. Where a captain rules, the sailors obey.

We have always look at both sides of the quarter. When we analyze with honesty all these laws and government programs, we will always see that the ones affected, are the people who work. All the people who work and have health insurance, this law doesn't benefit them, on the contraire, like it has been mentioned, it affects them except for the coverage of the previous conditions, which is the only thing to benefit everyone. For this reason, this law is being rejected. It benefits less than 30% of the citizens and harms more than 60%. For example, a father whose child is in college and still depends on his parents, this law benefits them; but a father whose son has finished high school and is between 18 to 26 years of age, and is already working, this law doesn't favor them at all. This is causing their kids to learn to be maintained, and not being adult men and women. This isn't strange for me since it is precisely what identifies the Democrats. I don't think you need to be an expert to know the percent of children that after finishing High School, continue to College, and how many finishes school. This is without counting the ones that never finish. This is one of the reasons the working class is opposed to this law. I ask myself, why does the government want to make me give my children health insurance? Since they are already working, and on the other hand I can't claim them in my taxes at the end of the year, because for taxes the government states they are adults, it doesn't matter if I maintain them or not, I am not allowed to claim them.

Like I said, no liberal press will report this. Besides this, not even the members of Congress truly knew the law they were signing, since even Nancy Pelosi, the leader of the Democratic Party, expressed that they had to first sign the law, to later find out what it was about. These types of decisions indicate to me that this woman, as well as other members of Congress, are a bunch of incompetents that should retire their positions which are too big for them, and go home and play with their grandchildren, instead of gearing the country to bankruptcy. Now they are faced with a big number of regulations and mandates that they do not want. For example, that contraceptives and condoms are included to prevent pregnancies. It's even incredible to hear Democratic members of congress express themselves when interviewed by the press, that they weren't aware of these mandates when they signed the law. These are the things that must be analyzed with honesty and transparency. Imagine where we have gotten, that the taxpayers, in the name of health, we are obligated to pay so that women can have sex and avoid getting pregnant. The nerve is so big, that therefore the Democrats accuse the Republicans of declaring war with women, because the Republicans are against these mandates. Of course, it is cheaper to give them contraceptives than having to maintain them, but it shouldn't be any of the two. It is the same government, mainly democrats, the ones who have created this problematic, that each day instead of correcting it, they want to continue. Now not only do we have to pay for housing, food stamps, health insurance, cell phones, and many other benefits, we must pay for contraceptives and condoms. There will come the day where they will legislate that we must pay taxes so that they can buy them cars, so they can move around freely under what they call, social justice. All these realities mentioned are worrisome. A big number of citizens are not aware or don't know, and as a result continue to choose these types of legislators, that in a matter of time will take the country to bankruptcy. The following is what an expert in economy predicts.

THE COUNTRY'S DEBT

The 13th of April 2013, a worldwide economy expert has predicted the falling of Wall Street in 2008, the decrease in sales and the purchase of homes, the falling of Freddie Mac and Fannie Mae, the bankruptcy of General Motors, the economic crisis that the United States of America faced, has done the following forecast. He forecasts a financial crisis worse than 2008; he forecasts that the Federal Government as well as some states will have to shut down their functions temporarily, the closing of banks and companies, and a new drop of at least 40% of Wall Street investments. He predicts that millions will lose their investments on bank accounts and pensions, and even the benefits of Social Security are in danger. He also predicts that the American dollar will lose a lot of value. This prognostic is based on the immense debt that the United State of America is in. As an expert in economy, he says that if the Federal government charges each citizen in the country 100% of the taxes, we will still fall short to the debt by trillions of dollars. This reminds me of another expert in economy, that when the debt was 14 trillion, it was bigger than if they had wasted a million dollars daily from the birth of Christ until now. This was when the debt was 14 trillion. The debt today is of 28 trillion and it's expected to pass the 30 trillion. The truth is that this debt is bigger, than the united debt of the European countries together. It's the biggest debt on the planet earth. To pay the debt of 28 trillion, it's costing an interest rate of

500 billion a year, and it's expected that for 2030 it reaches a trillion dollars annually, what I believe is the beginning of the end; and if the Democrats take the power, it will be the end.

President Obama and the Democrats thought that by manipulating the interests and giving the famous billion-dollar loans to private companies, it was going to stimulate the economy, but those have not been the results. Quite the opposite, the money they have loaned and wasted, is money they have also borrowed from countries like China and Japan, to who the country owes billions of dollars. According to the prognostics, very soon the cost of maintaining this debt will be declared incalculable, and impossible to pay, despite stimulating the economy. And how it has already been informed, even if they charge 100% of taxes to the citizens, and all the money from the multimillionaires of the country, they wouldn't reach paying the interests of the debt.

Up to 2013, President Obama had increased the rich their taxes in more than a trillion and a half dollars (1.6); this is combination with the expenses of the country, didn't come close to cover the interests and expenses of the country in one month. This indicates clearly, the amount of money that the Federal government is wasting annually. The wore part of this is that President Obama, like many of the Democrats, inform that it is not true that they are wasting money excessively. I don't understand what they call, giving out 8.4 billion dollars in loans to private companies, to build electric cars, like is the case of Fisker, that were assigned 200 million dollars and they went into bankruptcy. This is apart from the billions of dollars they have given out to solar companies, like the famous Solyndra, that they gave a 500-million-dollar loan from the money from taxpayers, and declared bankruptcy, and all that money was wasted. These are just a few examples, because the billions of dollars they have given out to private companies and cities, has no name. If taxpayers realized this or if they analyzed the millions of dollars being wasted in luxurious vacation trips, and several expenses, at the taxpayers' expense, they would have already done a revolution. In a report that I heard from a credible source, the wife of President Obama on a vacation she had in Spain during 2012, wasted in 5 days, half a million dollars from taxpayers' money. Vice president Biden wasted one million dollars in a few days when he visited Europe. As informed, the limousine alone that transported them from place-to-place cost 300 thousand dollars.

President Obama when He was running for his first term, called President George W. Bush an antipatriotic, because in eight years of his presidency, he added a debt to the country of 5 trillion dollars. In fact, 4 trillion dollars that were owed went up to 9 trillion. In their own words they had debited the future generations by thousands of dollars each. He was more specific saying that the children of our children, before they are born, are already in debt. Nevertheless, in four years he took the debt from 9 trillion to 17 trillion; and it's expected that before he finishes his second mandate, the debt should go past 23 trillion, without considering the interests and other things mentioned. Precisely that was what happened with the cost of Obama Care, with more than a trillion dollars.

There are still people today that use the excuse that President Obama inherited an economic disaster, for what was previously mentioned, reason why he hasn't been able to achieve good results. In my opinion, people who still at this stage use this excuse, is simply due to their political postures already mentioned, or from a lack of knowledge, or they don't have the capacity of analyzing certain truths. For example, when President Carter passed the presidency to President Reagan, he gave him a disaster three times worse than what Obama received. The country's economy was a total disaster, unemployment was on the ground, the interests of loans were 18%, almost at a war with Iran, since they had 52 American hostages that the incompetent Carter couldn't liberate. What about the lines to get 4 gallons of gas every other day? Communism covering the Caribbean and Central America. The war in El Salvador, invasion of Granada and Santo Domingo. People that know, remember that during the Carter presidency, the whole world lost respect for this country. It was a lot more the problems that Reagan inherited. Nevertheless, in his first three years of presidency, he corrected everything that has been mentioned. He was reelected leaving a prosperous country and the world once again, including Russia, respected us. Of course, Reagan before becoming president, had a good resume. He was an exalted governor of one of the biggest states of the nation regarding population. With all honesty, what was Obama's resume? I don't think he would have run a McDonalds. His resume in reference to a coach was of a neighborhood organizer. Having studies does not make you capable to run a leader nation of the world. These were just the results of his incompetence, because although he was a Senator when he was elected president, he only had the experience for one year, because he invested the other year in his political campaign.

All experts know that all loans are based on the value of the guarantees of what we possess. The same way, the loans of the country are based on the country's treasury. When the debt goes beyond those guarantees or treasury, it's time to pay or lose everything. That is why two years ago, the country's credit came down for the first time in history, from a Triple Score (AAA) to a double (AA). And if they don't impose the famous abduction, we would already be in the classification (A). There still exists the possibility that this classification will go down, precisely because of the uncontrolled expenses that the country has. The experts in economy assure that even if we take drastic measures, we are in a difficult economic situation that we can't overcome. Can it be overcome? I believe so. But imagine the cuts and measures that the government would have to take. If with the famous abduction cut, that is only an 87 billion dollar cut, they are screaming, imagine if they see themselves having to cut 4 or 5 trillion? We must inform that these 87 billion cuts, only covers the cost of two days in the country. There is a great possibility that the same thing happened to the banks Freddie Mac and Fannie Mae with the Housing Market, could happen to the country at government level. There are cities already that have declared bankruptcy. I believe it's time for both parties, and even the citizens, to stop pointing blame or looking for who is guilty, because when we analyze this with honesty, what has been happening for so many years, many many, many, years, there are millions at blame. I mention it because there are millions of tricksters here that for many years have cost trillion of dollars to taxpayers in the programs that the same government has fabricated. I can assure you that many of them are the ones blaming President Bush. I think that they still haven't come up with means to stop this money waste, because the President, many legislators and citizens believe the country is too big and powerful to come down. They forget that before this world potency, there were 5 potencies that thought the same. That was the case of Spain who is almost in bankruptcy and with one of the worst unemployment in the world. This is also the case of the old Greek Empire. In any case, if they don't do it, it's a matter of time when all the citizens will be affected, without any remedy on something we can already see coming.

We must acknowledge that President Clinton was an architect in fixing the economy and creating jobs through the reforms already mentioned. But, years later one of the reforms brought catastrophic results of which the press never informed correctly. It is regarding the amendment of the Community Investment Act, signed by President Carter in the year 1977. In this amendment the system that approves loans, the use of credit cards, and everything that had to do with the purchase and sales based on credit. For his years of presidency, this amendment gave results of trillion dollars in all types of purchases and sales, including home purchases, cars, credit cards, everything. That is how they took the country from a deficit to an economic status positive. But this amendment, years later brought the following catastrophic results. The press conveniently and maliciously, with political interests has never informed the following. Why the malicious intent? To understand what happened to the country in 2008, there are three factors that contributed to the problematic that we went through for 8 years. The first was the falling of Wall Street, another was the petroleum increase, that of course when petroleum goes up, we not only see it at the pump, but in everything impacted by the economy, from what we eat to what we pay in mortgage and rent. The other was the market drop in housing. What is clear is that everything has a beginning. All of this did not occur overnight like the people that keep blaming GWB think. It took several years, and that is why today we have the results we have, that occurred during the administration of President Clinton. The following, without a doubt was what caused the Mortgage market to collapse.

In the mid 70's years, only the rich were able to buy properties and have businesses. In the year 1977, ex-President Carter signed the law previously mentioned, Community Investment Act, to favor the middle class that had a good credit and a down payment. The banks could not deny them the opportunity for a loan to obtain a home if they had the income to sustain the purchase. It was a good law that started a movement to see middle class families obtain their homes and have small businesses. In the year 1995, this law was emended by ex-president Bill Clinton and in the new law, a loan can't be denied to anyone, even if they had bad credit. What was allowed was for the bank to charge interest based on the credit. The better the credit, the better the interest the person would have. The more affected the credit was, the higher interest rate payment was applied. From this point on is that you see that no matter the credit, everyone could purchase. What did this do? Open the door to all types of deceiving and fraud that many people are facing today. This amendment created many buyers, and by having so many buyers, property costs increased. It also created many buyers in other types of businesses. I know a man that has a hardware store, and he told me that during the presidency of Clinton, after amending the law, it has been the years where he has made the most in all his time in the business. He has been in business for a long time and explains that not before or after, he had ever sold so much to people paying with credit cards. In my case, my house was valued at $212,000 when we bought it in 2002. In the years 2005 and 2006, because there were so many buyers interested in the houses in this community, the houses were selling between $450,000 to $500,000 in a matter of days. Now that there aren't as many buyers, it has gone back down to $210,000 and they are not selling. These realities will not be reported by the liberal press. The report that properties have lost their value, but they don't report the real reasons why. Of course, as I mentioned, nothing that damages the Democrats they will report. The mistake of the George Bush administration and the Congress was not to stop what was already happening. The money that the banks were lending was at risk since they were lending the money to irresponsible people, that were also being approved by companies that were taking advantage and based on fraudulent information were being qualified, and it would come to a point where they wouldn't be able to pay it. I should also clarify that President Bush tried to put a stop to all this, but Democratic Congressman Barney Frank from the state of Massachusetts, who was the Affordable Housing leader, had already imposed regulations to banks during the Clinton administration, forcing the banks to increase the requisites making it harder to qualify for a loan from 30% to 50%. To add to the problem, the homeowners took advantage of the equity and took out loans based on the new values. When home values went down, and the buyers diminished, many people chose to give up their properties because they couldn't pay it. To many of these loaners, it is easy to return the properties because many of them have already enjoyed those loans in some matter. For example, I went in to get a job done in my car on the breaks, and while I waited there was a guy who was having a conversation with another person and was telling him that he bought a house for $160,000. The value went up to over $260,000. With that equity he received a loan of $100,000. Now, the house's value came back down to $160,000. With the 100,000 he bought a pickup truck that cost him $60,000. He was referring to a truck that was a double wheeler in the back and he was buying new wheels because, he was planning to return the house and move to Mexico. In a joking manner he said he was already enjoying the money; the bank could take back the house. Now, since there weren't many buyers, the home value was on the ground. It doesn't take a genius to know that the less buyers there are, the less values the properties have. These are the things that the liberal press will not report, because it is not convenient to them. But this was a true reality.

I have heard people relate the drop of the market to the war at Iraq. The market drop has nothing to do with the war, but rather with the finance market of the private sector, companies, investors, and the movement of buy and sell. The war in Iraq is in part responsible for the country's deficit. The deficit has to do with the money of the taxpayers. It also has to do with the wrongly management of funds by the government. That lack of knowledge is what take many people to have the wrong ideas and concepts of other people, like for example President Bush. The truth is that others have created laws that have contributed to these serious problems that we have faced. Nevertheless, the only thing you hear is that President Bush is the only responsible for the whole disaster. Of course, it is mere politics from politicians and their allies, the liberal press.

THE WAR IN IRAK

I should inform you that ex-president Bill Clinton said textually in 1998, that Iraq had destruction weapons, and that Sedan Hussein represented a danger not just for that region, it also was for the rest of the world. He even went further to say that sooner or later he needed to be removed from the power. Later, around the year 2000, his wife, Senator Hillary Clinton expressed the same thing, as did Senator John Karry, John Eduard, and many other Democrats. After they had to face the facts of the war, they informed that President Bush had lied to them to go to war. This for me are little kid excuses that only ignorant people believe. Although I know that they did it for political convenience.

Those that do not know the word of God get tied up easily with the opinions and information provided by the liberal press. Their only objective is to create news to satisfy their own egos, as well as their preferred Democrats, and make money. Of Course, they have also become a school that doctrine their viewers to their liberal ideology. There is a very powerful reason that explain these facts. When President George W. Bush won the elections in his first term, the Democrats couldn't accept the defeat. They went to court and even appealed to the Supreme Court, seeking for the court to give them the victory. Despite all of what they did, they didn't achieve what they wanted. In his second term, according to the press, in the polls he didn't have probabilities of wining those elections. This because of all the problems with the war in Iraq, his popularity indicated a guaranteed defeat. Not even the Republican party believed he would be able to win his second term. Nevertheless, we can say it was a miracle he was able to win the elections. Without the knowledge of the word of God, we can't understand these things. Christians at all levels, both laymen and officials, and even Pastors, preach one thing, but live another. For example, we preach that without God's permission, not even a leaf from a tree moves. We preach that God is who allows for everything to happen; that God places and removes governments, and effectively, the word of God in Romans 13:1 says: *There is no authority except that which God has established.* To understand the war in Iraq, and the things that will come, you must know the word of God. When we read the word of God, we find that for Israel to arrive at the promise land, they had to fight certain wars that were allowed by God. In 1 Chronicles 19-18, the Syrians went to war with Israel, and David killed 4,000 Syrians with his General. Today people are scandalized because 4,000 soldiers died in the war with Iraq. I don't want to be misunderstood, nobody wants to see anyone die, including myself. But since the beginning of the world, to survive, it has been through wars. The ambition of many governments and some organizations have always been by trying to dominate the world and establish power. In the year 1941, Japan attacked the United States, without this company provoking them and as a result, there was a war in which hundreds of thousands of people died. In those times, nobody was complaining like they are now. In the book of Daniel, chapter 8 verses 1-12, we see that God gave Daniel a vision when he was in front of the Ulai River. Daniel says he raised his eyes and saw a ram get up and do what it wanted. It hurt with his horns from the west, the north and south, and that no one was able to stand in front of the ram because of the power it had.

This ram reminds me of the ex-dictator of Iraq, Saddam Hussein, since that was exactly what he had been doing. Also, this vision that was given to Daniel was in the same geographical area, because the river that is mentioned here is precisely in Iraq, that in the Old Testament of the bible was called Babylonia. I am not saying that it was referring to him, but it does have his same characteristics. It could be possible that God allowed for the story to repeat itself. Daniel says that when he considered this, a male goat came from the west over the face of the earth. What calls my attention is that it came over the face of the earth, without touching earth. Meaning we can say that it came from the air, and through the air only planes travel. The male goat reminds me of President George W. Bush, that I am also not saying it refers to him. But, when we analyze what happened in Iraq and this vision, the characters and facts are the same. I remind you that planes had not been invented until after the year 1930. And, if there weren't any planes, there weren't any pilots either. In Revelations 18:17 tells us, that in the events that were happening, there were already pilots, and these do have a relation to the vision that was given to Daniel. This male goat took the power away from the Ram, and threw him on earth, and no one could liberate the Ram from his power. This was exactly what happened to the ram of Saddam Hussein; the male goat defeated him. And it was from a cave on earth that he took him out. Another thing that I can't understand today is the lack of humanity that people are living by. Like I said before, the opposes of the war, in majority hate Bush, for his decision on the war with Iraq. But, for the Iraqi people, he is an idol, and they even constructed a monument in his memory. Specially the women, who were liberated from the submission and abuse they were receiving. For those that do not know, I will inform the following. In a US TV channel (Discover Channel), they presented a reporter that sneaked into Iraq as a person that was studying and discovering archeology stuff. But what he truly did was record films of thousands of Iraqis, that were killed and mutilated with the so called, nuclear arming or chemicals in the hands of the dictator Saddam Hussein. They found common tombs with thousands of cadavers, and people incarcerated in underground jails dying from hunger and screaming for justice. Maybe you will say: what does this have to do with the decision to go to war? Only God knows that who knows all things. Psalms 48:8 says that even the storm winds execute his word. In this same Iraq was that King Nebuchadnezzar rebelled against God and became big in power. So, God used Daniel to inform him that he would be punished by Him, taking away his reign and the power, and took him to eat grass like the cattle in the field. That is why I believe you shouldn't get tangled by the press. Satan will use any means to keep people entertained and fooled. The press and the Internet are powerful instruments the devil is using to damage and hurt the same Christians. The press in this country are almost all liberals that are against the Christian principles, and they only favor those that think like them. The Iraqis lived under a criminal regimen that did with their people whatever they pleased, committing all types of crimes and abuse. It is easy for us to have an opinion and go to bed with a full stomach, live without those fears, and live without the oppression of a Dictator of that magnitude. But like I already mentioned, God is the only one that knows until when He allows the injustice and the abuse, and who He allows or use so that His will be fulfilled. In this case I do believe He used President Bush to liberate these people that lived under oppression. Another thing that I can't understand is the lack of humanity in which people live today. Like I said before, the opposers to the war, mostly hate President Bush for his decision on the war against Iraq. Nevertheless, for the Iraqis, mainly the women, he was an idol for the reasons I already mentioned.

THE OPRESSION OF THE CUBAN PEOPLE

The same thing that happened with Iraq, can happen with the Cuban people that are only 90 miles away from our coasts, and there is a regimen that commits the same abuses and run overs that have the Cubans going through hunger and misery, while no one in power dares to say or do something in protest. Quite the opposite, Congressmen like my own compatriot Serrano, Mr. Rangel, and others from the Democratic party, as well as many citizens including Christians, think that President Bush is a despicable person to them, but the Castro brothers are duo of good people. Time will be a witness, even if it takes many years to give ex-President Bush the reason, that he did the correct thing of eliminating that assassin dictator, that can no longer continue to terrorize people, or threatening the world, especially Israel. Of course, when we fill our mouths with opinions, we forget about these important things. When we go to the word of God, we see how God through wars, placed and removed governments, all because they did what was wrong in His eyes. On the other hand, my honest advice the Cuban people is for them to pray united to God and not to the virgin or the saints. I believe with all due respect, that these truths can bring them good results, that no statue, or no saint can hear them. Idolatry is an abomination to God. It's the fourth Commandment.

Exodus 20:4-5, 4 "You shall not make for yourself an image in the form of anything in heaven above or on the earth beneath or in the waters below. 5 You shall not bow down to them or worship them; for I, the Lord your God, am a jealous God, punishing the children for the sin of the parents to the third and fourth generation of those who hate me,

I mention this because although not all Cubans are idolaters, I believe there are many Cubans that depend on saints made by hands mentioned. Elizabeth, the mother of John the Baptist, in the book of Luke, chapter 1 verse 42, refereeing to the Virgin Mary said: Blessed are you amongst all women and blessed is the fruit of your womb. Verse 48 says that all generations will call her blessed. To be the woman who God chose to bring to the world the one and only Savior of the world, the Son of God, is to be blessed by all generations. Understanding she was a young and pure virgin when God chose her. Mary after got married to Joseph and had other children according to the stories that are found in the New Testament. In the same book, chapter 2, verse 48, is an evidence after Mary told Jesus after searching for him, your father and I were searching in anguish, referring to Joseph.

Acts 1:14, They all joined together constantly in prayer, along with the women and Mary the mother of Jesus, and with his brothers.

For me you will always be blessed, but nothing else. She, as do the rest of the humans, die. Nowhere in the bible does it say or imply that she should be adored. Just as those that have died, she can no longer hear us. That she is in the glory of God, there is no doubt, but the book of Ecclesiastes 9:5-6 says:

For the living know that they will die, but the dead know nothing;
they have no further reward, and even their name is forgotten.
⁶ Their love, their hate and their jealousy have long since vanished;
never again will they have a part in anything that happens under the sun.

This is why no saint, or no one after they have died can hear us. There are people that assure they communicate with the dead, and it's possible that they hear voices or see things. But this is product of other things that have to do with satanic influences or in other cases psychological. Idolatry is a sin and abomination to Jehovah, for God does not listen to the prayers of the idolaters. We can't generalize because there are those that pray and are heard. My point is that God doesn't listen to everyone like people think. I am clear that in His word we find his admonitions to pray without ceasing, for us to pray and pleadings, pray for the sick, and many verses that exhort us to pray. We also find a lot of word that confirms that God does not listen to people who disobey. David in Psalms 66:18 said:

If I had cherished sin in my heart, the Lord would not have listened; I am not talking about being perfect,

We have to live truly separated for God if we want God to listen and answer our prayers. For example, the person that abuses or treats wrong his wife, his prayers are not heard.

1 Peter 3:7, Husbands, in the same way be considerate as you live with your wives and treat them with respect as the weaker partner and as heirs with you of the gracious gift of life, so that nothing will hinder your prayers.

Let's imagine the following, we are liars, slanderous, cursing, practicing sin in many ways, and after, present ourselves before the Almighty God to talk to him. As we know not just anyone can go speak to Queen Isabel. Those that are allowed to speak with her, have to learn several manners, including their own family. If that is for Queen Isabel, imagine how clean and good spiritual behavior we have to have to present ourselves to the King of Kings and Lord of Lords, to speak to him. Its impossible to ask the Lord through saints, that according to His word is an abomination to God. God never hears those prayers. It's the reason why many people have needs and they pray to God, but He doesn't answer. For God to hear us we need to pay a price, and it's called obedience. That is why in John 15:7 says:

> *If you remain in me and my words remain in you, ask whatever you wish, and it will be done for you.*

Will our prayers be heard? This is just in reference in asking for our needs and sickness, Because for bigger things like the authority to rebuke evil things, there needs to be a consecration and cleanliness in prayer and fasting with the Lord, a lot more profound. We can see it clear in Mark 9, when the disciples weren't able to rebuke an evil spirit from a boy. When they asked Jesus, why they had prayed but weren't able to rebuke, Christ answered the following way: this gender doesn't go out unless with fasting and prayer. Profoundly I believe that the religiosity game is coming to an end. The only way God listens to us is when we repent, we ask for forgiveness, and we obey his commandments. Maybe you have never heard it, but the constant disobedience is a sin for which the wrath of God comes down. Colossians 3:5-10,

> *⁵ Put to death, therefore, whatever belongs to your earthly nature: sexual immorality, impurity, lust, evil desires, and*

greed, which is idolatry. ⁶ Because of these, the wrath of God is coming.[a] ⁷ You used to walk in these ways, in the life you once lived. ⁸ But now you must also rid yourselves of all such things as these: anger, rage, malice, slander, and filthy language from your lips. ⁹ Do not lie to each other, since you have taken off your old self with its practices ¹⁰ and have put on the new self, which is being renewed in knowledge in the image of its Creator.

This is not a critic. Quite the opposite, I love the Cuban people, and with all my heart I want to see them free and happy. I am sure that if they set aside their idolatry and pray to God with faith, they will be heard, and Cuba will become free. Matthew 9:18 says:

While he was saying this, a synagogue leader came and knelt before him and said, "My daughter has just died. But come and put your hand on her, and she will live."

We have to pray with belief and without any doubts, Jesus speaking to his disciples said in Matthew 21:21:

Jesus replied, "Truly I tell you, if you have faith and do not doubt, not only can you do what was done to the fig tree, but also you can say to this mountain, 'Go, throw yourself into the sea,' and it will be done.

You shouldn't believe or trust in images, in men, but only in God. He knows the means and has the power to create any miracle, like the miracle of the Iraqis. The devil shakes before God's presence, as do mountains. So, wouldn't the Castro's and their clique shake when the Almighty confronts them? On the other hand, I ask myself; where is the love for our neighbor that the word of God teaches us? Since the government today governs based on public opinion, people should raise their voices and tell the world, that in the century we are in, we will not tolerate all this injustice. The word of God tells the people that create injustice laws and abuse people the following: Isaiah 10: 1-4

> *Woe to those who make unjust laws, to those who issue oppressive decrees, ² to deprive the poor of their rights and withhold justice from the oppressed of my people,*
> *making widows their prey and robbing the fatherless.*
> *What will you do on the day of reckoning when disaster comes from afar? To whom will you run for help? Where will you leave your riches? ⁴ Nothing will remain but to cringe among the captives*
> *or fall among the slain. Yet for all this, his anger is not turned away, his hand is still upraised.*

THE WAR BETWEEN THE MUSLIMS AND HUMANITY

People think that what happened on 9/11 was something isolated and that it will not happen again. What I am about to speak on might be surprising. On Mr. Schnittsh's Talk Show, on Fox News on 6.10 am and from 3-6pm, they interviewed an extremist Muslim, that according to the presentation that was done to him, he is a teacher of the Koran, and is an expert in the doctrine and beliefs of this religion. Mr. Schnittsh asked him the following questions: Can you explain to us why Muslim men, women and children place bombs on their bodies and are willing to go out and die while killing others? How could you send men to detour airplanes that caused the death of over 3,000 lives knowing they would die as well? What explanation can you offer that would justify how Muslims in your country were celebrating, women, men, and children, when the towers came down and many people were killed? His answers were the following, (and I text verbatim): To be able to understand the reasons that justify everything that you have said, you have to know the following: God gave Abraham the laws by which he wanted people to live by. The people did not want to follow or live by them. So, God eliminated what was established and gave Moses new laws. They are the ones Moses wrote on stone. Time went by and people did not want to follow the laws that God gave Moses either. When

123

God saw that people were not obeying the laws, He had given Moses, He eliminated those laws and gave new ones to the Prophet Jesus Christ. After 507 years after He gave those laws to Jesus Christ, which are the laws that are in the Bible and are the laws that Christians live by today; He eliminated those laws from the bible and gave new laws to the Prophet Muhammad. The prophet kept receiving new laws and are the laws that are in the Koran. Those are the laws that tare valid for a period of 23 years. Since God has not called any other prophet, these are the laws we need to live by. The Bible and the laws of Moses expired, and now we live by the laws of the Koran. Under the Koran laws, Ala who you call God, instructed him that all of humanity must convert to Muslims after a determined time that has already expired as well. All those who do not convert to Muslims, would turn into demons and devils and that they would have to go after them and kill them. According to his explanation, all those that die for the cause of killing demons and devils, which is what we are to them, and die, goes straight to heaven, and is received in the arms of Muhammad. Now you might understand the reason they celebrate every time they kill demons and devils and those who sacrificed themselves went to heaven.

These are the reasons why the President of Iran has stated that Israel must be removed from the map. Israel is the only country of the Middle East that is not Muslim. These are the reasons why the extremist Muslims consider them demons and devils that must die. That is why the president of Iran has stated they need to continue creating the nuclear weapons, because according to him, no one can stop him from creating these weapons. He has expressed that Ala has given him instructions to erase Israel from the map. The president of Iran has said that no one, absolutely no one, can persuade him from building these nuclear weapons, because they have to listen to Ala and not any man. His reasons are religious and not political. That is why the Muslim extremists call everything they do and these terrorists acts, a holy war. The Muslims declare it. That is why we see the attempts are not just against the American people. It is more dangerous to see so many politicians place their personal interests ahead of the security of the citizens, who are the ones that elected them, like the treaty that Obama and John Kerry did with Iran. President Trump was allowed by God, not just so that the capital of Israel would go back to Jerusalem, but this treaty with Iran had to be eliminated. For now, Israel needs the protection of the USA, since it's the country God chose to protect it. For those that have not noticed, I believe it is not a coincidence that in the center of JER (USA) LEM are the letters of USA. United States of America. I also believe that taking out the troops from the Middle East, will bring a bigger conflict. When Israel sees itself much less protected, before Iran tries to eliminate them from the map, they will have to react, and the conflict will be worse. I also know that what God has established will be fulfilled, and these acts have to take place. I have always said and sustain that the day that all these things happen, the Democrats will be in the power here in the United States, because history talks about the acts and positions, for example the war with Iraq. I believe that the time for all things to end has arrived as said the Archangel Gabriel to Daniel. I want to make something clear; United States of America doesn't appear in the prophetic mentioning's of the end of times. As per some of the biggest theologians, the ones that do appear are Russia and Rome together with the Vatican and of course, many cities from the Middle East; but this will be at the end of time. Before this happens, there are many things that need to happen. I am referring to who will be in the power here In our country, that for now leads the world power, and will be the ones to let them happen. What I am sure about is that, according to history and by the political postures, is that it will be the Democrats. For example, in the times we currently live, it's more than visible to see a group of citizens have obtained control over the government, destroying the country, taking away the power of the police, the governors, violating the laws, declaring parts of the cities as autonomous, and demanding the police be eliminated. Cities like New York and others dominated by Democrats, have honored their demands. I am going to give you undeniable evidence that the Democrats will be responsible of the destruction of the United States of America. In the history of humanity, since the beginning, we can see many people did what was wrong in the eyes of the Lord, and as a result they were destroyed, burned, like the case of Sodom and Gomorra. Other cities were swallowed by the ground. The following are two of many examples.

> *Numbers 16:23-33, 23 Then the Lord said to Moses, 24 "Say to the assembly, 'Move away from the tents of Korah, Dathan and Abiram.'"*
>
> *25 Moses got up and went to Dathan and Abiram, and the elders of Israel followed him. 26 He warned the assembly, "Move back from the tents of these wicked men! Do not touch anything belonging to them, or you will be swept away because of all their sins." 27 So they moved away from the tents of Korah, Dathan and Abiram. Dathan and Abiram had come out and were standing with their wives, children, and little ones at the entrances to their tents.*
>
> *28 Then Moses said, "This is how you will know that the Lord has sent me to do all these things and that it was not my idea: 29 If these men die a natural death and suffer the fate of all mankind, then the Lord has not sent me. 30 But if the Lord brings about something totally new, and the earth opens its mouth and swallows them, with everything*

that belongs to them, and they go down alive into the realm of the dead, then you will know that these men have treated the Lord with contempt."

31 As soon as he finished saying all this, the ground under them split apart 32 and the earth opened its mouth and swallowed them and their households, and all those associated with Korah, together with their possessions. 33 They went down alive into the realm of the dead, with everything they owned; the earth closed over them, and they perished and were gone from the community.

Leviticus 16: 14-28, 14 He is to take some of the bull's blood and with his finger sprinkle it on the front of the atonement cover; then he shall sprinkle some of it with his finger seven times before the atonement cover.

15 "He shall then slaughter the goat for the sin offering for the people and take its blood behind the curtain and do with it as he did with the bull's blood: He shall sprinkle it on the atonement cover and in front of it. 16 In this way he will make atonement for the Most Holy Place because of the uncleanness and rebellion of the Israelites, whatever their sins have been. He is to do the same for the tent of meeting, which is among them in the midst of their uncleanness. 17 No one is to be in the tent of meeting from the time Aaron goes in to make atonement in the Most Holy Place until he comes out, having made atonement for himself, his household and the whole community of Israel.

18 "Then he shall come out to the altar that is before the Lord and make atonement for it. He shall take some of the bull's blood and some of the goat's blood and put it on all the horns of the altar. 19 He shall sprinkle some of the blood on it with his finger seven times to cleanse it and to consecrate it from the uncleanness of the Israelites.

20 "When Aaron has finished making atonement for the Most Holy Place, the tent of meeting and the altar, he shall bring forward the live goat. 21 He is to lay both hands on the head of the live goat and confess over it all the wickedness and rebellion of the Israelites—all their sins— and put them on the goat's head. He shall send the goat away into the wilderness in the care of someone appointed for the task. 22 The goat will carry on itself all their sins to a remote place; and the man shall release it in the wilderness.

23 "Then Aaron is to go into the tent of meeting and take off the linen garments he put on before he entered the Most Holy Place, and he is to leave them there. 24 He shall bathe himself with water in the sanctuary area and put on his regular garments. Then he shall come out and sacrifice the burnt offering for himself and the burnt offering for the people, to make atonement for himself and for the people. 25 He shall also burn the fat of the sin offering on the altar.

26 "The man who releases the goat as a scapegoat must wash his clothes and bathe himself with water; afterward he may come into the camp. 27 The bull and the goat for the sin offerings, whose blood was brought into the Most Holy Place to make atonement, must be taken outside the camp; their hides, flesh and intestines are to be burned up. 28 The man who burns them must wash his clothes and bathe himself with water; afterward he may come into the camp.

I don't think that a further explication is needed of all the barbarous things and sins that the Democrats have been committing in opposition of what God has established. Since a few decades, under the famous separation of Church and State, they have stepped on the divine commandments. Everything began in 1947 the following manner.

THE SEPARATON OF CHURCH AND STATE

In 1802, President Thomas Jefferson, because of some meddling of the Federal Government against a Baptist Church in the state of Connecticut, through a letter they asked the Congress to legislate a law so that there could be a separation of the church and state. Clearly their intentions were to protect the practice of religion and their doctrines that are clear in the amendment that precisely the Congress did base on this petition. All of this changed when a liberal judge, Hugo Black, in the year 1947 interpreted that the intention of President Jefferson was to place a wall between the government and the church. This interpretation that until this day is under discussion, has been the base of the liberal Democrats to place the mentioned demands, and all other demands they have placed after this one. For example, they have placed many demands, which all have been granted, like remove crosses from certain places, paintings, forbid Christmas trees in governmental offices, including saying Merry Christmas, and even the mandates to the religious institutions of providing birth control pills to the women under the Obama Care law. Since the year 1947 the liberals in this country have been fighting for all Christian principles to be taken out. In the year 1962, the liberal and atheist woman, from the state of Texas, Madelyn Murray O'Hair, collected thousands of signatures and went to court demanding for the Bible and prayers to be taken out of the school. The court that was composed mainly by liberal judges, voted 6 to 1 in favor of Mrs. O'Hair's petition.

In another demand in January 2005 by Kay Stanley, a lawyer that also works in Real Estate, demanded in Houston, Texas, that the Bible be removed from the court rooms in the Harris County, of the state. The court composed by a liberal judge, granted her petition and the Bible was removed. In an appeal to this demand by local officials, the Circuit Court of the United States in the city of New Orleans, Louisiana, with the majority of liberal judges, on August 25 of the same year, sustained the same votes 8-1. And as we all know, we no longer swear in court with the hand on the bible. There were other demands like the families of Hyde Park in the state of New York, that placed a similar demand as Mrs. O'Hair in the year 1963, and another demand in the same year by Mr. Ed Schempp in Philadelphia, PA. This last one was so that the reading of the bible would be prohibited in schools. Organizations like the ones already mentioned that place these demands, if these demands go to Superior Court, and the Superior Court rules in their favor, even if they are local or state, they become national laws that affect all states. This was the case of the last two demands that were mentioned. The Supreme Court joined these two demands, voting 6-1 in favor of the demanders converting the law into a national law. That is why you see in all of the trials the lawyers, fiscals, and the judges use decisions from other courthouses as guides for their decisions.

The liberals have accomplished removing prayer from schools, and it's forbidden to talk about the word of God, not even in the hallways. What is allowed is to teach about sex from an early age. The Bible is not allowed, but they can give out birth control pills to students, motivating them to have sex freely without the risk of getting pregnant. There are even schools that have approved that they can give out birth control to the students, without having to tell the parents. Even worse, we have come to the point that they don't need to be informed that their young girls are getting an abortion. The liberals will continue to fight against all Christian principles. Everyone knows that the Democrats is the one to approve all of these ideals. They were the ones that approved all of these to happen; the removal of the Bible of the court and now they want to remove, amongst other things the "In God We Trust" of the currency. During many years the liberals had tried to legalize abortions and the laws to authorize them. They have been able to do it every time they dominate the Congress and the Senate. The following is the quotes of the history of abortions.

THE HISTORY OF ABORTIONS IN THE UNITED STATES OF AMERICA

(QUOTE) Before the independence of the United States there were barely any laws regarding induced abortions and its penalty applying the common law that basically established that abortion was acceptable and legal if it was practiced before the mother felt the fetus (quickening). After the Independence, many different laws appeared in the decade of 1820. 1821 in Connecticut legislating on the abortion supplies to the pharmaceuticals and in New York penalizing the induced abortions.

Many of the first feminists, amongst them Susan B. Anthony and Elizabeth Cady Stanton, were opposed to abortions since they considered it to be the ending of a series of aggressions to the women and to the lack of true independence, that they felt needed to be corrected. Then a woman could reject sexual relations with her husband, of which unwanted pregnancies and abortions derive. There were no laws that protected women of rapes from their husbands, and women of low resources found themselves without the least independence for a divorce or sexual relations. To legalize abortions were for some of the first feminists, resolve a problem without modifying the cause.

During the decade of 1860 the legislation increased penalizing and criminalizing abortions. In 1900 abortions were illegal in many states although there were some that included special causes for an abortion. Generally, these were to protect the life of the woman, or pregnancies due to rape or incest. Despite the penalization, abortions continued into the 20th century, making the practice unsafe because it was considered illegal. In many cases the woman's life was at risk and could lead to death, like the case of Gerri Santoro of Connecticut In 1964.

Before the sentencing in the case of Roe vs. Wade, there were exceptions to the forbiddance of abortions in at least 10 states by rape, incest, and risk of the mother.

Approval of birth control methods.

(Thank God that President Trump took them to Superior Court, and they ruled in his favor that citizens do not pay taxes so they could hand out birth control and condoms to people so they can go out and have sex freely, at the cost of taxpayers).

Principle Article: Case Griswold vs. Connecticut

In 1965, the decision, 7 votes in favor and 2 against of the Supreme Court of the United States in the case of Griswold vs. Connecticut, sentenced that the Constitution of the United States protected the rights to privacy (Privacy Laws of the United States) and therefore, declared invalid the laws of the different states that violate the *marital privacy rights* that guarantees the access and the administration of birth control.

1n 1965, the ACOG-American Congress of Obstetricians and Gynecologists, assumed the position defended by Bent Boving in 1959 that considered conception started in the implantation of the embryo and not at when the fecundation is produced. This medical repot changed the category of some of the birth control methods that until then, were considered abortion methods when they acted before the implantation of the embryo in the uterus. In 2015 58% of the Americans supported abortions reaching its highest point in the last two years based on a survey by The Associated Press, that reveals an apparent increase in support between the Democrats and Republicans equally, during the last year.

Case Roe vs. Wade 1970-1973

Principle Article: Case Roe vs. Wade

In 1970, the lawyers Linda Coffee and Sarah Weddington, presented a demand in Texas representing Norma L. McCorvey ("Jane Roe"), claiming the right of induced abortion by rape. Although the district attorney of the County in Dallas, Texas, Mr. Henry Wade, who represented the state of Texas was opposed to the abortion. In the end, the district court ruled in favor of Jane Roe, but without establishing changes in the legislation of abortion included in the United States.

"Jane Roe" gave birth to her daughter and placed her in adoption while the case was not yet decided.

Fourteenth Amendment; Right to Privacy as the Ultimate Right

Principle Article: Fourteenth Amendment to the Constitution of the United States

The case was appealed on many occasions until it arrived at the Supreme Court of the United States, that in their resolution on the 22nd of January 1973, established that women have the right to the free choice, known as the right to the privacy or intimacy, that protected the decision of taking a pregnancy or not, to a full term. According to the sentencing the right of privacy derives itself from the clause of *due process*, of the *fourteenth Amendment* of the *Constitution of the United States of America*. The decision provoked a change in all of the state and federal laws that proscribed or retrained abortions and that were contraire to the new decision.

Case Doe vs. Bolton 1973

Principle Article: Case Doe vs. Bolton

If the essential content of the Supreme Court sentencing case Roe vs. Wade was that *abortion must be allowed to any women, no matter the reason, until the point where the fetus transforms into viable, in other words, it is able to live outside the maternal uterus, without artificial help,* the sentencing of the case Doe vs. Bolton, published on the same 22nd of January 1973, established that *induced abortion should be legal when necessary to protect the woman's health.*

From the beginning of abortions in 1973, over 60 million babies have been executed. The question I ask is the following: Do you really believe that after all I have informed, and what I will continue to inform, God is not going to bring punishment to a rebel and criminal humanity? Spiritually, without a doubt, we are wrong. I am not talking about being religious, I am talking about how we have a totally different and reasoning behavior.

THE WARNINGS DUE TO SIN

The way I see things, the Democrats are in the wrong in almost all aspects. Both in politics and in the decisions they make. It's undeniable the behavior of many legislators as well as a vast majority of people. As it has been informed in 2015, 58% of Americans approve abortion. Today I am sure that percentage is a lot higher. As well as same sex marriage. I believe that for many the following information can be taken as fanaticism or old school, it's necessary to give the same warnings that God gave us. If we really believe in God and what the Bible teaches, we have to pay attention to the following information. One of the warnings that the Lord left us in His word in Matthew 24 and Luke 21, was that at the end of times, which we are currently, because of evil and disobedience, plagues like COVID-19 would occur. When God gave Daniel the revelation, at the end of chapter 12 He ends up telling him:

"As for you, go your way till the end. You will rest, and then at the end of the days you will rise to receive your allotted inheritance."

There are many that do not believe that evil and sin will not bring punishment from God, but that is not what the bible teaches. The God from the Old Testament is the same God of the New Testament, and as the one cited by the book of Revelations, that narrate the punishments that will come because of sin. Even though there are

Christians that think that just because we are living under the grace, God won't bring punishment. That is a big misconception. The punishments from sin begin in this season, and end in the judgments of Revelations. When people pass their limits and insist rebelliously to irritate God, they have always received punishments from God. For example, in the book of numbers, chapter 16:20-33:

20 The LORD said to Moses and Aaron, 21 "Separate yourselves from this assembly so I can put an end to them at once."

22 But Moses and Aaron fell face down and cried out, "O God, the God who gives breath to all living things, will you be angry with the entire assembly when only one man sins?"

23 Then the LORD said to Moses, 24 "Say to the assembly, 'Move away from the tents of Korah, Dathan and Abiram.'"

25 Moses got up and went to Dathan and Abiram, and the elders of Israel followed him. 26 He warned the assembly, "Move back from the tents of these wicked men! Do not touch anything belonging to them, or you will be swept away because of all their sins." 27 So they moved away from the tents of Korah, Dathan and Abiram. Dathan and Abiram had come out and were standing with their wives, children, and little ones at the entrances to their tents.

28 Then Moses said, "This is how you will know that the LORD has sent me to do all these things and that it was not my idea: 29 If these men die a natural death and suffer the fate of all mankind, then the LORD has not sent me. 30 But if the LORD brings about something totally new, and the earth opens its mouth and swallows them, with everything that belongs to them, and they go down alive into the realm of the dead, then you will know that these men have treated the LORD with contempt."

[31] As soon as he finished saying all this, the ground under them split apart [32] and the earth opened its mouth and swallowed them and their households, and all those associated with Korah, together with their possessions. [33] They went down alive into the realm of the dead, with everything they owned; the earth closed over them, and they perished and were gone from the community.

Jeremiah 7:25-28, From the time your ancestors left Egypt until now, day after day, again and again I sent you my servants the prophets. 26 But they did not listen to me or pay attention. They were stiff-necked and did more evil than their ancestors.'

27 "When you tell them all this, they will not listen to you; when you call to them, they will not answer. 28 Therefore say to them, 'This is the nation that has not obeyed the Lord its God or responded to correction. Truth has perished; it has vanished from their lips.

Jeremiah 8:4-8,

[4] "Say to them, 'This is what the LORD says:

"'When people fall down, do they not get up?
When someone turns away, do they not return?
[5] Why then have these people turned away?
Why does Jerusalem always turn away?
They cling to deceit;
they refuse to return.

⁶I have listened attentively,
but they do not say what is right.
None of them repent of their wickedness,
saying, "What have I done?"
Each pursues their own course
like a horse charging into battle.
⁷Even the stork in the sky
knows her appointed seasons,
and the dove, the swift and the thrush
observe the time of their migration.
But my people do not know
the requirements of the LORD.

⁸ "'How can you say, "We are wise,
for we have the law of the LORD,"
when actually the lying pen of the scribes
has handled it falsely?

Jeremiah 8:12, Are they ashamed of their detestable
conduct? No, they have no shame at all; they do not even
know how to blush. So, they will fall among the fallen;
they will be brought down when they are punished, says
the Lord.

Jeremiah 9:1-11, Oh, that my head were a spring of
water
and my eyes a fountain of tears!
I would weep day and night
for the slain of my people.
2 Oh, that I had in the desert
a lodging place for travelers,
so that I might leave my people
and go away from them;

for they are all adulterers,
a crowd of unfaithful people.

3 "They make ready their tongue
like a bow, to shoot lies;
it is not by truth
that they triumph[b] in the land.
They go from one sin to another;
they do not acknowledge me,"
declares the Lord.
4 "Beware of your friends;
do not trust anyone in your clan.
For every one of them is a deceiver,[c]
and every friend a slanderer.
5 Friend deceives friend,
and no one speaks the truth.
They have taught their tongues to lie;
they weary themselves with sinning.
6 You[d] live in the midst of deception;
in their deceit they refuse to acknowledge me,"
declares the Lord.

7 Therefore this is what the Lord Almighty says:

"See, I will refine and test them,
for what else can I do
because of the sin of my people?
8 Their tongue is a deadly arrow;
it speaks deceitfully.
With their mouths they all speak cordially to their
neighbors,
but in their hearts they set traps for them.
9 Should I not punish them for this?"

declares the Lord.
"Should I not avenge myself
  on such a nation as this?"

10 I will weep and wail for the mountains
  and take up a lament concerning the wilderness
grasslands.
They are desolate and untraveled,
  and the lowing of cattle is not heard.
The birds have all fled
  and the animals are gone.

11 "I will make Jerusalem a heap of ruins,
  a haunt of jackals;
and I will lay waste the towns of Judah
  so no one can live there."

Psalms 11:4-6, The Lord is in his holy temple; the Lord is
on his heavenly throne.
He observes everyone on earth; his eyes examine them.
The Lord examines the righteous, but the wicked, those
who love violence, he hates with a passion.
6 On the wicked he will rain fiery coals and burning
sulfur; a scorching wind will be their lot.

Proverbs 1:24-26, But since you refuse to listen when I call
and no one pays attention when I stretch out my hand,
since you disregard all my advice and do not accept my
rebuke, I in turn will laugh when disaster strikes you; I will
mock when calamity overtakes you—

I believe that the information that the Bible gives is important. First, because it's the word of God and second, because everything that the Bible has informed has been accomplished down to the last letter. I also believe that it is very important for humanity to understand that because of the prophecies and things to come, we are already living the end of times, in reference to a series of events that will take us to the prophecies of Revelations. What is happening is just the beginning of sufferings. We have all heard there will come an antichrist. It was the government that was revealed to Daniel 536 years before Christ. The same one that was revealed to John in Revelations 95 years after Christ. He analyzes the theology and the events that are taking place. He knows that the mentioned government will soon take power.

WORLD ORDER

We all know that for decades, globally they have tried to establish a world order. The world order is being backed up and financed not only by the biggest world leaders, but by companies and billionaires in the world. For example, people are not paying attention, and we are living in a time where people have an itching to hear, as 2nd of Timothy 4:3 says. There is an organization at a global level known as the Illuminati Pyramid. (The word illuminati derive from Latin meaning illuminated). This organization began functioning in the 18th century. Its foundation was in Germany. Today as per the available information it is composed by the richest, most powerful people in the world, political leaders, and by the renown 13 families of Royal Lineage. Their objective is to create a global government that can govern all countries. They also have the purpose of reducing dramatically the world population. This organization today is very well organized. We also have to pay close attention to the well-established groups of the council of 3, council of 5, council of 7, council of 9, council of 13, council of 33, the Grand Druid Council, Committee of 300, and the Committee of the 500. The way they want to govern the world is by creating a governmental system in which there is a global leader, elected by the leaders of those organizations. Their plan is to create a system with just one currency. Create a system of armed forces. Create a system of global police. Their purpose is to create new laws for all countries and enforce these laws through this system. According to the information, people who refuse to the laws they have established will be accused of rebellion and their punishment will be a capital penalty. The Illuminati organization is receiving the support of multimillionaire companies all over the world. Here in the United States, there are over 200 companies that are in support. For years they have been infiltrated different Universities all over the world, including the US. They have institutes and organizations working directly for them. According to some information, they have the so called Progresso's; these are the liberals from the Democrats that call themselves infiltrated Progresso's in the same government here in the US, working so that this government can be established. In my opinion these Democratic Progresso's mentioned, have already achieved to reduce the population here in the US by more than 50 million habitants, with the law they legislated granting the abortion crimes. If not, we would have already been close to 400 million of habitants.

This world order, in the Trump administration, has served as an obstacle since Trump has opposed to this order even before he was elected as president. I say by God's will, since God is the one who knows when He will allow it to happen. I believe this is the reason why the Democrats have done everything in their power to take him out since day one, after he was elected. But I believe that as soon as it is established, they will name a world leader, and that leader will be the antichrist that was revealed to Daniel and John. This governor is the one identified in the bible the following way. In the book of Daniel 8:23, it presents us the antichrist like a prince that is to come. Daniel 9:26 like the bleak; 9:27 like a despicable man; 2nd of Thessalonians like the son of doom. Revelations 11:7 as the beast. According to theologians and professors in eschatology, they say he will be a Jew and he will surge from a gentle nation, possibly Syria or Greece.

I consider everything to be already organized. It is nothing else but the government of the antichrist. Like we all know, the Bible teaches us that the antichrist will be the one to rule the world. During his government, one currency will be used. It will be the government that will establish new laws. It will also be the government that globally will control the world in politics, religion, and commerce. They will be the ones to kill all those opposed or that do not obey their laws. In our faces today, there are a number of organizations that are using naïve young people that have no clue what they will demand, like the elimination of the police.

What is already happening I consider it to be the beginning of the prophecies of the final times given to Daniel in chapter 7 to 12.

THE ANTICHRIST GOVERNMENT

Daniel 7: 1-8, In the first year of Belshazzar king of Babylon, Daniel had a dream, and visions passed through his mind as he was lying in bed. He wrote down the substance of his dream.

2 Daniel said: "In my vision at night I looked, and there before me were the four winds of heaven churning up the great sea. 3 Four great beasts, each different from the others, came up out of the sea.

4 "The first was like a lion, and it had the wings of an eagle. I watched until its wings were torn off and it was lifted from the ground so that it stood on two feet like a human being, and the mind of a human was given to it.

5 "And there before me was a second beast, which looked like a bear. It was raised up on one of its sides, and it had three ribs in its mouth between its teeth. It was told, 'Get up and eat your fill of flesh!'

6 "After that, I looked, and there before me was another beast, one that looked like a leopard. And on its back, it had four wings like those of a bird. This beast had four heads, and it was given authority to rule.

7 "After that, in my vision at night I looked, and there before me was a fourth beast—terrifying and frightening and very powerful. It had large iron teeth; it crushed and devoured its victims and trampled underfoot whatever was left. It was different from all the former beasts, and it had ten horns.

8 "While I was thinking about the horns, there before me was another horn, a little one, which came up among them; and three of the first horns were uprooted before it. This horn had eyes like the eyes of a human being and a mouth that spoke boastfully.

In this same chapter 7 of Daniel, from verse 15 to 28, it describes who these beasts are and what all of this means.

15 "I, Daniel, was troubled in spirit, and the visions that passed through my mind disturbed me. 16 I approached one of those standing there and asked him the meaning of all this.

"So, he told me and gave me the interpretation of these things: 17 'The four great beasts are four kings that will rise from the earth. 18 But the holy people of the Most High will receive the kingdom and will possess it forever—yes, for ever and ever.'

19 "Then I wanted to know the meaning of the fourth beast, which was different from all the others and most terrifying, with its iron teeth and bronze claws—the beast that crushed and devoured its victims and trampled underfoot whatever was left. 20 I also wanted to know about the ten horns on its head and about the other horn that came up, before which three of them fell—the horn that looked more imposing than the others and that had eyes and a mouth that spoke boastfully. [21] As I watched, this horn was waging war against the holy people and defeating them, [22] until the Ancient of Days came and pronounced judgment in favor of the holy people of the Most High, and the time came when they possessed the kingdom.

[23] *"He gave me this explanation: 'The fourth beast is a fourth kingdom that will appear on earth. It will be different from all the other kingdoms and will devour the whole earth, trampling it down and crushing it. [24] The ten horns are ten kings who will come from this kingdom. After them another king will arise, different from the earlier ones; he will subdue three kings. [25] He will speak against the Most High and oppress his holy people and try to change the set times and the laws. The holy people will be delivered into his hands for a time, times and half a time.[a]*

[26] *"'But the court will sit, and his power will be taken away and completely destroyed forever. [27] Then the sovereignty, power and greatness of all the kingdoms under heaven will be handed over to the holy people of the Most High. His kingdom will be an everlasting kingdom, and all rulers will worship and obey him.'*

[28] *"This is the end of the matter. I, Daniel, was deeply troubled by my thoughts, and my face turned pale, but I kept the matter to myself."*

This is the same vision that was given to John in Revelations 13: the two beasts.

The dragon[a] stood on the shore of the sea. And I saw a beast coming out of the sea. It had ten horns and seven heads, with ten crowns on its horns, and on each head a blasphemous name. 2 The beast I saw resembled a leopard but had feet like those of a bear and a mouth like that of a lion. The dragon gave the beast his power and his throne and great authority. 3 One of the heads of the beast seemed to have had a fatal wound, but the fatal wound had been healed. The whole world was filled with wonder and followed the beast. 4 People worshiped the dragon because he had given authority to the beast, and they also worshiped the beast and asked, "Who is like the beast? Who can wage war against it?"

5 The beast was given a mouth to utter proud words and blasphemies and to exercise its authority for forty-two months. 6 It opened its mouth to blaspheme God, and to slander his name and his dwelling place and those who live in heaven. 7 It was given power to wage war against God's holy people and to conquer them. And it was given authority over every tribe, people, language and nation. 8 All inhabitants of the earth will worship the beast—all whose names have not been written in the Lamb's book of life, the Lamb who was slain from the creation of the world.[b]

9 Whoever has ears, let them hear.

10 "If anyone is to go into captivity,
   into captivity they will go.
If anyone is to be killed[c] with the sword,
   with the sword they will be killed."[d]

This calls for patient endurance and faithfulness on the part of God's people.

The Beast out of the Earth

11 Then I saw a second beast, coming out of the earth. It had two horns like a lamb, but it spoke like a dragon. 12 It exercised all the authority of the first beast on its behalf and made the earth and its inhabitants worship the first beast, whose fatal wound had been healed. 13 And it performed great signs, even causing fire to come down from heaven to the earth in full view of the people. 14 Because of the signs it was given power to perform on behalf of the first beast, it deceived the inhabitants of the earth. It ordered them to set up an image in honor of the beast who was wounded by the sword and yet lived. 15 The second beast was given power to give breath to the image of the first beast, so that the image could speak and cause all who refused to worship the image to be killed. 16 It also forced all people, great and small, rich and poor, free and slave, to receive a mark on their right hands or on their foreheads, 17 so that they could not buy or sell unless they had the mark, which is the name of the beast or the number of its name.

18 This calls for wisdom. Let the person who has insight calculate the number of the beast, for it is the number of a man.[e] That number is 666.

The 10 horns that had 10 headbands mean a confederation of nations that will be governed by the government of the antichrist. He will go even further. He will be a commercial political leader and religious, he will control the commerce and the world economy, where no one will be able to buy or sell, but those that have the seal or mark of the beast as previously mentioned.

In the vision Gabriel tells Daniel in the book of Daniel 8:17:

As he came near the place where I was standing, I was terrified and feel prostrate. "Son of man," he said to me, "understand that the vision concerns the time of the end."

Daniel 12: 8-9, I heard, but I did not understand. So, I asked, "My lord, what will the outcome of all this be?"

9 He replied, "Go your way, Daniel, because the words are rolled up and sealed until the time of the end.

In the prophecy of Daniel 11, there are two events. The first one is about the war between the kings of the north and the south, from verse one to twenty. The second in reference to the antichrist that shall come, verses 21-45. The wars mentioned in verses 1-20 have already taken place. The kings of the north, Greece; and the kings of the south, Syria, Egypt, and Babylonia. In these wars prophesized by Daniel, also participated the powerful King Ahasuerus, who reigned from India to Ethiopia from 486-465 BC. With all of his power, he sustained a war with Greece, but Greece defeated him. These wars lasted until King Epiphanes took power typifying the antichrist, year 163 BC. The Romans were also involved who had already become a world potency and through a war they conquered Syria in the year 63 BC, as well as Israel. Nevertheless, in reference to the wars of the north and south regarding the oriental dominance, there has been a parenthesis for more than 2,200 years; but the war between the north and the south will come again with the famous was known as Armageddon. This will be the fulfillment of the second part of this prophecy, with the beginning of the antichrist of verses 21-45. This war will be against the glorious land of Israel, Egypt, and Syria. Possibly as well as with the Confederation of Arab Countries against the countries of the confines of the north, which is Russia and many allies that will be directed by prince Gog, who is the antichrist as per Ezekiel, 38.

The prophecies are being fulfilled in our faces and people are not noticing. That is why, those who know the word of God, we know that soon the government of the antichrist will take place. For example, the prophecy of Damascus, city of Syria, was written thousands of years ago, and it was just accomplished. For the first time in their history, the troops, and the bombings of American and Russian airplanes, converted her to ruins. Isaiah 17:1-3:

> *A prophecy against Damascus:*
>
> *"See, Damascus will no longer be a city*
> *but will become a heap of ruins.*
> *² The cities of Aroer will be deserted*
> *and left to flocks, which will lie down,*
> *with no one to make them afraid.*
> *³ The fortified city will disappear from Ephraim,*
> *and royal power from Damascus;*
> *the remnant of Aram will be*
> *like the glory of the Israelites,"*
> *declares the LORD Almighty.*

In Revelations 13, John saw a beast come up from the sea, that meant an umpire. Theologians and experts in eschatology affirm that it is about the Roman Umpire that will reign again. Revelations 17:8-18,

> *⁸ The beast, which you saw, once was, now is not, and yet will come up out of the Abyss and go to its destruction. The inhabitants of the earth whose names have not been written in the book of life from the creation of the world will be astonished when they see the beast, because it once was, now is not, and yet will come.*
>
> *⁹ "This calls for a mind with wisdom. The seven heads are seven hills on which the woman sits. ¹⁰ They are also seven*

kings. Five have fallen, one is, the other has not yet come; but when he does come, he must remain for only a little while. ¹¹ The beast who once was, and now is not, is an eighth king. He belongs to the seven and is going to his destruction.

¹² "The ten horns you saw are ten kings who have not yet received a kingdom, but who for one hour will receive authority as kings along with the beast. ¹³ They have one purpose and will give their power and authority to the beast. ¹⁴ They will wage war against the Lamb, but the Lamb will triumph over them because he is Lord of lords and King of kings—and with him will be his called, chosen and faithful followers."

¹⁵ Then the angel said to me, "The waters you saw, where the prostitute sits, are peoples, multitudes, nations and languages. ¹⁶ The beast and the ten horns you saw will hate the prostitute. They will bring her to ruin and leave her naked; they will eat her flesh and burn her with fire. ¹⁷ For God has put it into their hearts to accomplish his purpose by agreeing to hand over to the beast their royal authority, until God's words are fulfilled. ¹⁸ The woman you saw is the great city that rules over the kings of the earth."

This city is Rome, the commercial and religious Babylonia. This is the only city in the world that is two cities in one. The Vatican and the commercial Rome. The Vatican that is adorned how Revelations 18:16-19 says and:

and cry out:

"'Woe! Woe to you, great city,
*   dressed in fine linen, purple and scarlet,*

and glittering with gold, precious stones and pearls!
[17] In one hour such great wealth has been brought to ruin!'

"Every sea captain, and all who travel by ship, the sailors, and all who earn their living from the sea, will stand far off. [18] When they see the smoke of her burning, they will exclaim, 'Was there ever a city like this great city?' [19] They will throw dust on their heads, and with weeping and mourning cry out:

"'Woe! Woe to you, great city,
*   where all who had ships on the sea*
*   became rich through her wealth!*
In one hour she has been brought to ruin!'

Is it not the Vatican the city adorned with these riches? The antichrist will be the political, commercial, military, and religious leader. As you can see, this will finish as it started 2,083 years in reference to the taking of Israel by the Roman Umpire. This is why I reaffirm that God is who allows those who govern, for He has a plan for what needs to happen, happen. This is why Romans 13:1 says:

Let everyone be subject to the governing authorities, for there is no authority except that which God has established. The authorities that exist have been established by God.

A clear example is as follows: Tel Aviv had replaced Jerusalem as the capital of Israel. In Luke 21:24 there is a prophecy that will be fulfilled during the government of the antichrist. Jerusalem will be trodden down by the gentiles, until the time of the gentiles is fulfilled. (The 7 years of the great tribulation. For this to be fulfilled, Jerusalem had to become capital again. The past three

159

administrations, they promised to restore Jerusalem as capital, but they never did. It is clear to me that God used this president to fulfill this prophecy. This is one of the reasons I insist to different Christians to submit themselves to what has been ordered in the word of God and not to what is said in the liberal press. The important thing is not what already happened, but what is about to happen. John talked about seven kings. When they gave him the prophecy, five had already fallen: Egypt, Assyria, Babylonia, Mid Persia, and Greece. The sixth was the Roman that lasted until 1453 DC. The next, that still hasn't come is the Roman (that was, is not, but will return). The eighth that John mentions, Revelations 17, is the government of the antichrist. The one that Daniel 11:21 says that to this despicable man will not be given the honor of the kingdom but will come without warning and will take the reins with compliments. And as I mentioned USA does not appear in any final event; what indicates that it will lose the power. Based on history and the actions taken, there are two reasons that will take the country to the loss of power. The first has already been established, they have turned their back to God. More or less from God's friends to God's enemies. The second involves the economy and the debt. In both cases the responsible are the Democrats. All that needs to happen is for God to allow them back on the power, so they can invest trillions in the climate change, and give everything free, and goodbye America.

The false prophet

The antichrist will use a false prophet that will help open the door I order to accomplish everything that has been mentioned. It will be someone like John the Baptist that opened the way for our Savior Jesus Christ. This will be the second beast that John saw in Revelations 13:11-18 and:

¹¹ Then I saw a second beast, coming out of the earth. It had two horns like a lamb, but it spoke like a dragon. ¹² It exercised all the authority of the first beast on its behalf and made the earth and its inhabitants worship the first beast, whose fatal wound had been healed. ¹³ And it performed great signs, even causing fire to come down from heaven to the earth in full view of the people. ¹⁴ Because of the signs it was given power to perform on behalf of the first beast, it deceived the inhabitants of the earth. It ordered them to set up an image in honor of the beast who was wounded by the sword and yet lived. ¹⁵ The second beast was given power to give breath to the image of the first beast, so that the image could speak and cause all who refused to worship the image to be killed. ¹⁶ It also forced all people, great and small, rich, and poor, free and slave, to receive a mark on their right hands or on their foreheads, ¹⁷ so that they could not buy or sell unless they had the mark, which is the name of the beast or the number of its name.

¹⁸ This calls for wisdom. Let the person who has insight calculate the number of the beast, for it is the number of a man.[a] That number is 666.

As we can see, he disguises himself as a lamb, but speaks like a dragon, but rather, he is not what he appears to be. The devil himself will give him the power and the same authority he has to do miracles and signs like making fire descend from the heavens, and everything else that has been written here. The question we all ask ourselves: Who is it? Some theologizes that are profoundly invested in this prophecy have stated it will be a Jew descendant from Palestine. Other theologians think it will be the Pope. There are two pastors in

YouTube that give the names of two priests who believe that the Pope is the antichrist. The truth is that no one can prove who it will be. Nevertheless, even though the following is not evidence, it is not assumptions and just leaves a big question mark. For example, the prophecy is regarding a global religious leader. Until now, the Pope is the only world religious leader. He has been creating many reforms to the church in which many Catholics, including priests, have not agreed. He has made anti-biblical comments such as: God will not send anyone to hell, not even atheists. He has taken a defensive excessive posture in the climate change, to which he has stated that we have sinned severely against the earth. But he has not mentioned that Obama and the Democrats have sinned greatly against God, marrying homosexuals, and killing 60 million babies. When he was visiting Washington DC during the presidency of President Obama, he emphatically treated this subject in which he left very clear the obligation of all the countries and people in the world to take actions to save the planet. I believe he appears not to know who is in control of the world and the universe. I already spoke clear on this matter and what the word of God states. Why is this important? Revelations 11:12 says: Then they heard a loud voice from heaven saying to them, "Come up here." And they went up to heaven in a cloud, while their enemies looked on. Note that it states that it makes the earth worship the beast, and not just the dwellers of the earth.

The following is a quote of a Catholic Reporter.

The concept of this Pope as a "Great Reformer"

The first is regarding the cultural war that anyone in the West know very well the conflict between the moral teachings of the church and the way we live today, the struggle of sexual ethics of the New Testament and if they should be revised or abandoned due to the post realities of sexual revolution. The Pope's plan is witty or misleading, depending on the point of view. Instead of changes formally the teachings of the church regarding divorce and new nuptials, marriage between the same sex, or euthanasia; changes officially impossible since they are far beyond the authority of his charge. The Vatican Francisco is undertaking a double action. One report says that the in last year, Pope Francisco added another change when he looked for a truce, not with a culture, but with a regimen: the communist government of China. Francisco wanted a deal with Beijing that would recon ciliate the underground Catholic church, faithful to Rome, with the "Patriotic" Catholic Church, dominated by the communists. If achieved, that reconciliation would require the church to explicitly give part of its authority to name bishops to the political bureau.

This information and more information that has not been informed, in sot evidence to confirm that the Pope can be the antichrist, but his expressions and reforms take me to believe what Revelations 13:11 already mentioned, he dresses like a lamb but speaks like a dragon.

All of these acts that have been prophesized have already had their accomplishments. And like I informed, the Unites States of America does not appear in any of the end of times prophecies. The United States of America will disappear and no longer be the world potency that is currently controlling the world. For years I have been reading books of Theologians on Eschatology and of profound studiers of the prophecies that will be taking place at the end of times. For example, the professor Kittin Silva, at the beginning of the 80's wrote a book on the prophecies of the book of Daniel and Revelations. He interpreted very detailed the acts of the prophecies given to Daniel and John. He also informed that Russia would fall and come back up again. As we know, everything has already happened. All of the strong theologians agree that United States of America will not appear in the acts of the end of times that will take place like for example, the Armageddon War; and the nations that will arise, mentioned by Daniel. They all concur that Russia is the country mentioned by the Bible in the prophecies as the country that will come from the north, to fight in the Armageddon war. If you haven't noticed, it has not also arisen, its already under agreement with United States, during the Obama administration, leading the war in Syria. But, once again I go back to what I have said about United States. For years I asked myself, how such a powerful country like this one would lose its power. Although the other 5 empires, who were powerful during their times, like Spain and Greece also fell. I believe that we are too focused in politics, and we have forgotten about the most important, the prophecies. That's why I want to give this information. I didn't think this would be possible and I believe that many think the same. But, after seeing how the liberal politicians have taken God out of everything, without a doubt, this country is already morally dead; killing over 50 million babies, allowing same sex marriage, and the long list of everything that has been mentioned. Society in general calls the good, bad; and the bad, good. The prophecy in Daniel 12 is already being fulfilled, where it also states that the ungodly, will act ungodliness. It is clear to me that things will only get worse. What is also clear to me is that the country is almost at a point of bankruptcy with a 30 trillion-dollar debt. They owe 28 trillion but are forecast at 30. The experts are already confirming that the debt will be greater than 30 trillion. When we add up all of the events of the country that are already many, there is no doubt that through the same government and a sick society, without the intervention of others, we will lose it all and our owners will be the Chinese and the Japanese. And since we will no longer have God's protection, it's only a matter of time for:

Goodbye America. That is how the biblical prophecies will be fulfilled. Now I see clearly why United States doesn't appear in the end of times.

Regarding the prophecies, nor the politicians or the press, know where we are standing. Someone at CNN says that Trump is not prepared or has a plan to stop the pandemic. My response to this comment is the following: Nor Trump or anyone else can detain what the Supreme God has established need to occur. The world domination belongs to God, and even though politicians and the press believe that the United States is very powerful, and could dominate and fix any situation, they are completely wrong. As proof that US will not be the leader, we can find it in Ezekiel 38 in the Armageddon war that Gog and Magog will fight; but rather Russia, who is the country that will come from the West against Israel. As we know in all wars, including the wars in the Middle East, US has been leading them. But, in this war US and the coalition of countries will not be present since God Himself will defeat them. Ezekiel 38:18-23:

> *18 This is what will happen in that day: When Gog attacks the land of Israel, my hot anger will be aroused, declares the Sovereign LORD. 19 In my zeal and fiery wrath I declare that at that time there shall be a great earthquake in the land of Israel. 20 The fish in the sea, the birds in the sky, the beasts of the field, every creature that moves along the ground, and all the people on the face of the earth will tremble at my presence. The mountains will be overturned, the cliffs will crumble, and every wall will fall to the ground. 21 I will summon a sword against Gog on all my mountains, declares the Sovereign LORD. Every man's sword will be against his brother. 22 I will execute judgment*

on him with plague and bloodshed; I will pour down torrents of rain, hailstones and burning sulfur on him and on his troops and on the many nations with him. ²³ And so I will show my greatness and my holiness, and I will make myself known in the sight of many nations. Then they will know that I am the LORD.'

The majority of people thing that Unites States is too powerful to disappear as the world power. That was the same thing the other 5 umpires thought, including the last one, Spain who owned everything, including states here in the US, like the case of the treaty Adams-Onis that Florida was granted in 1821. In 2008, in a report I mentioned that the Democrats would take this country to ruins. And in a matter of time, we would disappear as the leaders of the world. In our faces they are doing just that, and we have not noticed. Already morally and spiritually the Democrats have killed it. Now, they are working frenetically to lose the only thing that's left, the economic power. In their plans there are trillions of dollars assigned to stop the climate, something that belongs to God. Like I mentioned before, Daniel prophesized that the science would evolve, and we know it has been accomplished. Because of these advances, we believe we are powerful. Due to the scientific advances, we have a space station and the satellites that all work through it, computers, cellular phones, and everything else, depend on those advances. This also includes investments, buying and selling, everything. In the book of Obadiah, God states the following in reference to the advances by men.

"I will cut you down to size among the nations;
  you will be greatly despised.
³ You have been deceived by your own pride
  because you live in a rock fortress

and make your home high in the mountains.
'Who can ever reach us way up here?'
   you ask boastfully.
⁴ But even if you soar as high as eagles
   and build your nest among the stars,
I will bring you crashing down,"
   says the LORD.

Imagine if a virus has put a whole world on their knees, and the world economy has been shaken, millions without a job, depending on the government to send a check. I want to see if God gives a nudge to the spatial station and the satellites, and we go back to 1800, to see where the power we think we have, is gone, and how many checks will be sent. What has been written will occur. As a good citizen and Christian, I will not be part of the failure and total ruin of the country by voting for the enemies of God, the Democrats. Aside from all this, the Democrats have always had a cowardly posture in regard to making firm decisions to defend the country. What actions did the democrats take when they tried to run down the Twin Towers in 1993? According to credible information, years later the CIA informed President Clinton they had captured the criminal terrorist Bin Laden. The president ordered for him to be let go, in fear of the opinion of other countries, despite having the evidence that it was the hand of Bin Laden behind the attack in Africa of the American embassy. This is why there are people that think that because of that negligence is why the events on 9/11 occurred. We also have to refresh our minds and know who were in charge of this country, when the situation of the hostages in Iran occurred during the Carter administration; the attacks against the ship of soldiers that died due to a rocket bomb that was launched to the terrorists and the US Embassy in Africa that was taken down with bombs and hundreds of people died. The list of the attacks around the world that terrorists have done against this country, is very long. This is without the other acts they have done against other countries as well. The word of God doesn't tell us who will be in the power, but it is clear that there will come events against the Glorious Land that we all know is Israel. According to the word of God, that will be fulfilled, there will be conflicts in the cities of the Middle East and knowing that United States is the potency that is controlling the world, it is clear to me, that it will be the Democrats that will be in the power when all these events take place. The positions that the Republicans have assumed do not need to be mentioned, since I believe everyone knows the stories of what they have done when they have been in the power. Simply remember what President Regan did when Granada, the invasion in Santo Domingo, Nomar Kadaffi, the Berlin Walls, and the communism in Central America. The position that George Bush father and Bush son assumed regarding the Persian Gulf and Iraq. I don't want to go too back in history but, the Republican position has always been brave in difference to the Democrats with some exceptions. History books have narrated to us all of the events and the decisions that have been made in this country, and who have made them.

Without a doubt, those of us that are aware of the theology, that the biblical prophecies, in all sense, in science and in human behavior, are being fulfilled, as it was written that would occur. The fact that this information is rejected, and people don't believe, and think that it is fanatism, or decide to ignore these warnings, are part of the prophecies itself.

In reference to the events against the Glorious land of Israel, I inform you that the government of the antichrist will govern for seven years. During the first three and a half years, the Jew will obtain the peace that for thousands of years they have been seeking, and no one has been able to find. This governor, with God's permission will achieve the peace, and they will be fooled, because the Jew are still waiting for the Messiah, having rejected Jesus Christ as the Messiah. The first three and a half years of his rein, according to the prophecies, this governor will fool the world, and the world will adore him and pay petechial. After the three and a half years, for an additional three and a half years, God will allow him to lead the great tribulation, where everything that has been mentioned will occur; the way he will govern, eliminating the police, changing the laws and the army, one currency, the religion, and everyone will have the mark 666 in order to buy and sell.

During these three and a half years, the Jew will suffer death and persecution and will fall at the edge of the sword. They will be taken captive to all the nations of the world. Jerusalem will be destroyed by the gentiles, until the times of the gentiles is accomplished. Luke 21:24,

> *They will be killed by the sword or sent away as captives to all the nations of the world. And Jerusalem will be trampled down by the Gentiles until the period of the Gentiles comes to an end.*

Nothing of this is fanatism, it's the reality that the entire world will have to face. If you believe what happened during the flood with Noah, when he told humanity to repent and they did not believe him, until the flood came, and they all perished. The same thing will happen to this generation of nonbelievers. The following warning is the one that Christ left us in Matthew 24:29-51,

> *[29] "Immediately after the anguish of those days,*
>
> *the sun will be darkened,*
> *the moon will give no light,*

the stars will fall from the sky,

   and the powers in the heavens will be shaken.[a]

30 And then at last, the sign that the Son of Man is coming will appear in the heavens, and there will be deep mourning among all the peoples of the earth. And they will see the Son of Man coming on the clouds of heaven with power and great glory.[b] 31 And he will send out his angels with the mighty blast of a trumpet, and they will gather his chosen ones from all over the world[c]—from the farthest ends of the earth and heaven.

32 "Now learn a lesson from the fig tree. When its branches bud and its leaves begin to sprout, you know that summer is near. 33 In the same way, when you see all these things, you can know his return is very near, right at the door. 34 I tell you the truth, this generation[d] will not pass from the scene until all these things take place. 35 Heaven and earth will disappear, but my words will never disappear.

36 "However, no one knows the day or hour when these things will happen, not even the angels in heaven or the Son himself.[e] Only the Father knows.

37 "When the Son of Man returns, it will be like it was in Noah's day. 38 In those days before the flood, the people were enjoying banquets and parties and weddings right up to the time Noah entered his boat. 39 People didn't realize what was going to happen until the flood came and swept them all away. That is the way it will be when the Son of Man comes.

[40] "Two men will be working together in the field; one will be taken, the other left. [41] Two women will be grinding flour at the mill; one will be taken, the other left.

[42] "So you, too, must keep watch! For you don't know what day your Lord is coming. [43] Understand this: If a homeowner knew exactly when a burglar was coming, he would keep watch and not permit his house to be broken into. [44] You also must be ready all the time, for the Son of Man will come when least expected.

[45] "A faithful, sensible servant is one to whom the master can give the responsibility of managing his other household servants and feeding them. [46] If the master returns and finds that the servant has done a good job, there will be a reward. [47] I tell you the truth, the master will put that servant in charge of all he owns. [48] But what if the servant is evil and thinks, 'My master won't be back for a while,' [49] and he begins beating the other servants, partying, and getting drunk? [50] The master will return unannounced and unexpected, [51] and he will cut the servant to pieces and assign him a place with the hypocrites. In that place there will be weeping and gnashing of teeth.

I know there are many that didn't even believe that God existed. For example, a coworker once told me when we were taking a trip together to resolve an issue, that he didn't believe in God because no one, absolutely no one, had been able to prove to him that God existed. He continued to tell me, if you could prove that God is real, I will accept him. I truly believe you won't be able to since I have asked this same question to many people, and they haven't been able to do it. And I can't believe in something that I can't see. He asked me, have you been able to see God? My answer was simple. I don't have anything to prove to you because to believe in God, is an act of faith. You don't have faith, He isn't real to you, but that doesn't mean that He isn't real. The following information made me realize that He is real and that He does exist. Based on what he told me that he couldn't believe in something that he couldn't see, I asked him the following: You say you don't believe in things you can't see, but have you seen the wind? He answered no, but that he could feel it and he was also able to see the branches when they moved. My answer was that I also haven't seen God, but that I could feel Him when He touched me; and in the same way you see the branches move, I can see the sea roll back up in return, I can see Him in the plants, in the same trees that you see move and that have life. Men can fabricate plants and trees, but they can't give the life. They are plastic. I see Him in how He made the universe, the perfection of the plants, the sun, the moon, and all the celestial bodies. I see Him in the perfection of how He made humans. Just thinking about the way our organs work, and the purpose of each one, the perfection of the millions of neurons and cells, and how each has a function for the proper functioning of the body, including the movements and thoughts, the intelligence, in summary, everything we are and are able to do. There are some that believe we are the product of a bacteria and others of a monkey, but in none of the two cases they can speak on so much perfection. Overall, since I believe in Him, there have been difficult times in my life, and when I have prayed to Him in my pain, without a doubt I have seen his hand in my favor. Taking me out of situations that were impossible to defeat unless through His intervention. I saw my mother already evicted by doctors, and totally a paraplegic due to three brain strokes, she had lost her voice completely, as well as her right side, stranded 24-7 to her bed; and God lifted her up again. It was a miracle acknowledged by the doctors since they had sent her home to die. I ended up telling him I could finish the year so many things that can demonstrate the God exists, but that I believed that was more than enough. Not having any arguments for such a big reality, he admitted that for the first time, he had nothing to say.

I ask myself, since the beginning of humanity, people have always been more rebellious and chose to do what's wrong to the eyes of the Lord. They have received punishment after punishment, and instead of turning to God, they continue to do what's wrong. Why are they so evil?

It's time to meditate and analyze our behavior as intelligent beings, with a spirit given by God, which has a capacity of renovation. To each human being God has given him a spirit that is the fountain of life to man. The soul is the owner of this life and uses it, and through the body it is expressed. Animals have souls but not spirits. The spirit represents the most elevated nature of man and its related to the nature of his character. For example, if we let jealousy dominate us, or liars, slanderers, perverse, etc. We need forgiveness, and lord it over our spirit, also makes a spirit contrite and humiliated. Ezekiel 18:31 says,

> *Put all your rebellion behind you and find yourselves a new heart and a new spirit. For why should you die, O people of Israel?*

The soul is intelligent and the one that cheers the human body through the body senses and the organs. But the soul is sinful. That is the reason why there is a constant battle within us between doing the good or the bad. Galatians 5:17,

> *The sinful nature wants to do evil, which is just the opposite of what the Spirit wants. And the Spirit gives us desires that*

are the opposite of what the sinful nature desires. These two forces are constantly fighting each other, so you are not free to carry out your good intentions.

When we allow the soul to defeat the spirit with its sinful desires, we are in a state of death. We allow the bad attributes identify our spiritual state. It's time to seek God in spirit and truth, through the constant renovation, through the Holy Ghost since our spirit is not able to vivify itself. God gave us free will to choose between god and evil; and even though the Catholic church believes in purgatory as a means of purification and salvation, that is not what the word of God teaches us in 2 Corinthians 5:10,

For we must all stand before Christ to be judged. We will each receive whatever we deserve for the good or evil we have done in this earthly body.

Why these warnings? Like I have mentioned, God gave Daniel the prophecies of everything that would happen to the end of times. It is evident based on the recent events, and the fulfilled prophecies, that these times have begun.

CHRISTIANS AND POLITICS

Years ago, and still, the majority of pastors and religious leaders have maintained the belief, that the church shouldn't get mixed with politics. I too believe that the church should not get involved in politics. But I do think that pastors and religious leaders have the responsibility of educating the church on politicians. This is important because there have been politics that have arisen with an antichristian attack and legislating laws that are totally opposed to what God has established, and what the church teaches. All of this of course is works of Satan, like the bible says in 1 Peter 5:8,

> *Stay alert! Watch out for your great enemy, the devil. He prowls around like a roaring lion, looking for someone to devour.*

Spiritually we are in a war and the church is supposed to be part of God's army. So, as its said in 2 Timothy 2, we are soldiers in this army to fight against the enemy. In a war, a soldier that is not well prepared, or have a good armor, is a dead soldier. In an army, the leaders prepare well the soldiers in order to win the battles. In the church the leaders have the responsibility before God, to prepare well the soldiers for the attacks from politicians, that are being used by Satan to destroy the church.

Besides this, there are many organizations that have risen to fight against all Christian principles, and they are doing protests and placing pressure on the Legislators to obtain what they want and demand. All of these things that are happening are just part of the fulfillment of the biblical prophecies. Like the days of Noah in Genesis 6, people mock it and don't want to listen. But those of us that know the word of God, we know that soon the government of the antichrist will take its place. People's behavior and the governments around the world, especially the administration of President Obama, approving everything that goes against what God has established, as well as giving their backs to the people of Israel, are all clear indications of the fulfillment of the biblical prophecies. The results we see, all of these groups have rebelled against God and the church, and they have obtained what they want and demand. Of course, all of this will continue to happen. My biggest concern is that the majority of the churches are entertained and are allowing these attacks on behalf of corrupt people, while Christian people are not doing anything that is not singing hymns at the temple and remain in the same circle for weeks, months and years. Although our Lord Jesus Christ in Matthew 16:18 said: *the doors of hell will not prevail against the church.* He also said in 1 Timothy 6:12 you should fight the battle of faith. In the book of James, chapter 2, and verses 14 to 26, teaches us that we need to act, because faith without works is dead. A clear example of this is the following: In the year 2010, in the city of King, North Carolina, the organizations of ACLU and the Americans United For The Separation Of State and Church, demanded before the mayor John Carter and the commissioners of this city, that a Christian flag that had been raised for many years at the Veteran's Memorial Park Cemetery, be removed from this place. John Carter and the commissioners by the recommendation of the marshals, made the decision to grant the demand, and they removed the flag from this place. In this small town there seem to be many Christians because they demanded a meeting with the mayor and the commissioners, and they demanded the flag to be returned to its place. The mayor and his team refused to go back. The Christian citizens and residents of this town decided to act on the works of their faith, and they started to buy and raise Christians flags in all of their homes and commercial buildings. When these officials started to see the amount of people that were sending out a message through the flags, and how the results would look like the day of the elections, immediately they restored the flag, and placed it back in its place. This was the result of the works of their faith.

All of these organizations not only use the protests, they also use the internet, the phone, and the mail. And they contact the legislators and place pressure on them, telling them that if they don't legislate in their favor, they won't give them their vote. Christians are much more millions, and we could do the works, and use the same methods and at least taking them out of their positions with our vote. Even more important, I want to make sure that this type of politicians don't get into the power with our votes, since once they're in the power, the start to create laws that we later regret. A clear example is in the state of New York, How many Christians helped with their votes place all those Democratic legislators, with the help of two liberal republicans, to pass the law that allows marriage between the same sex? Some 20 years ago, anyone that would have prognoses that the actual government would have been not only supporting these types of behaviors, but also legislate laws granting these marriages, the response would have been, absolutely not. In that same way, there will come a day where you will go into the temple and find a sign saying this temple has been closed by orders of the government.

Since we are living in the middle of a fragile and delicate society, where everything that is said or is informed is because we discriminate or is fanatic. I want to make it clear, that my main purpose is to defend the DEVINE laws; laws that a true Christian should defend over any other law or criteria, here on earth. For these same reasons is why there are true Christians dying in the hands of ISIS terrorists, and other organizations around the world. I hope that it is understood that the same way we Christians must respect and not discriminate against anyone; we also have constitutional rights that should also be respected in the same way. The constitution of the United States, in the first amendment says: religions have the right to exercise and practice the doctrines of their religion. For example, according to the established law, homosexuals have a right to get married in this country. But, according to our constitutional rights already mentioned, they nor anyone can make us participate of their sin. I want to make it clear that discrimination is not the same as not wanting to participate. For example, if I have a business and I refuse to sell or serve them because they are homosexual, that is discrimination; and that is not what the Bible teaches. To participate is to approve their homosexuality, to be part of their activities like their weddings, protests, parties, etc. The bible teaches not to be participant with those that practice sin. This is not limited to people who practice homosexualism, it refers to all types of sin. For example, if someone is committing adultery and I am somehow helping that relationship, I am participating on that sin.

The Legislators, homosexuals, and all of the people who defend them, need to understand that because of our constitutional rights, they cannot force us to violate the Divine commandments that are in the word of God. These rights were clearly understood by the Fathers of our homeland that wrote it in the constitution. They should also understand that not being participants of their sins, does not make us discriminators. Further along I will quote what the bible states in this matter.

An organization that is being strongly used by Satan, are the atheists. Of course, this organization doesn't believe in God, and I can imagine that they don't believe in the devil, but they are responsible for many demands in the court against all Christian principles. Since they don't believe in God it is easy for them with their diabolic minds, to try to damage all the great work that the church has done. If we don't wake up to reality, it will be harder each day for the church to work and many people will be headed for perdition. Spiritually and mentally the world is living under perdition as in Sodom and Gomorrah. Even if many think the contraire, we need to counteract sin. That is the function and the job of the church and of every Christian.

The question that everyone asks is, what does politics have to do with Christians? Reading the content based on the word of God, we can see the connection. What is happening is so true, that the press informs that the elections have been lost because Christians refuse to go out to vote. After reading all of the story you can see that the church has nothing to do with politics, but politicians have a lot to do with the church; and Christians have a lot to do with the politicians. Everything I talk about is based on the word of God with the bible references.

THE CHURCH

For many years the church has been announcing the second coming of our Lord and Savior Jesus Christ. The bible says that not even the angels that are in heaven know the day or time when this will take place. But He left us signs and events that would happen so that we could be alert and aware. He left us the parable of the 10 virgins in Matthew 25 and 24 where it narrates the events that are taking place on earth. There will be hunger, war rumors, plagues, nations against nations, earthquakes, and many saying I am the Christ, just as the Reverend Jose Luis De Jesus Miranda has been announcing himself, from the cult growing in grace. In verse 11 of this chapter, it says that many false prophets will rise and fool many. Without a doubt we are living the end of times. We will be seeing signs in the skies, wars, hunger, plagues (virus), earthquakes, nations vs. nations, and big events here on earth. The Bible tells us in Hebrews 13:8, that Jesus is the same today, yesterday and all centuries. What that means is that His laws and doctrines, don't change. Among the requirements that we find throughout the bible, is that those He considers a son, He tells them to take their cross and follow Him (Mark 8:34). To take the cross means to live a life separated for Him and live by the commandments. 1 John 2:15-17:

> *Do not love this world nor the things it offers you, for when you love the world, you do not have the love of the Father*

in you. ¹⁶ For the world offers only a craving for physical pleasure, a craving for everything we see, and pride in our achievements and possessions. These are not from the Father but are from this world. ¹⁷ And this world is fading away, along with everything that people crave. But anyone who does what pleases God will live forever.

To last forever is an expression that doesn't have an end, because although many people think that when they die everything is over, there is a different reality. There is an eternity of glory with God, or a hell with the devil. For many, this can sound hard, but this is what the word of God tells us. Matthew 25: 43-46:

⁴³ I was a stranger, and you didn't invite me into your home. I was naked, and you didn't give me clothing. I was sick and in prison, and you didn't visit me.'

⁴⁴ "Then they will reply, 'Lord, when did we ever see you hungry or thirsty or a stranger or naked or sick or in prison, and not help you?'

⁴⁵ "And he will answer, 'I tell you the truth, when you refused to help the least of these my brothers and sisters, you were refusing to help me.'

⁴⁶ "And they will go away into eternal punishment, but the righteous will go into eternal life."

Many people think the world will end and that will be the last of it. The bible tells us that what will be over is the devil's reign, the fallen angels, and all spirits of darkness. I mention this because the world ending has been misinterpreted from the bible prophecies. Like the devil, all of those that have are not saved and have not accepted the sacrifice of the cross and has not lived a life according to the divine commandments, will also go to perdition and to the lake of fire and sulfur. Revelations 20:10,

> *Then the devil, who had deceived them, was thrown into the fiery lake of burning sulfur, joining the beast and the false prophet. There they will be tormented day and night forever and ever.*

In reference to people's destiny Matthew 18:8 says:

> *So, if your hand or foot causes you to sin, cut it off and throw it away. It's better to enter eternal life with only one hand or one foot than to be thrown into eternal fire with both of your hands and feet.*

> *Revelations 20:15, And anyone whose name was not found recorded in the Book of Life was thrown into the lake of fire.*

> *Matthew 5:21-22, "You have heard that our ancestors were told, 'You must not murder. If you commit murder, you are subject to judgment.' But I say, if you are even angry with someone, you are subject to judgment! If you call someone an idiot, you are in danger of being brought before the court. And if you curse someone, you are in danger of the fires of hell.*

> *Matthew 18:9, And if your eye causes you to sin, gouge it out and throw it away. It's better to enter eternal life with only one eye than to have two eyes and be thrown into the fire of hell.*

> *2 Peter 3:5-7, They deliberately forget that God made the heavens long ago by the word of his command, and he brought the earth out from the water and surrounded it*

with water. 6 Then he used the water to destroy the ancient world with a mighty flood. 7 And by the same word, the present heavens and earth have been stored up for fire. They are being kept for the day of judgment when ungodly people will be destroyed.

For people that think hell is just a place of torment, and this will be the place where they will spend eternity, I have to share the following information. Hell is the translation of Sheol in the Old Testament, also known as Hades in the New Testament. Hell is the waiting room; the holding place of those that have died without salvation, to be judged on Judgment day. Luke 16:22-26:

"Finally, the poor man died and was carried by the angels to sit beside Abraham at the heavenly banquet.[a] The rich man also died and was buried, 23 and he went to the place of the dead.[b] There, in torment, he saw Abraham in the far distance with Lazarus at his side.

24 "The rich man shouted, 'Father Abraham, have some pity! Send Lazarus over here to dip the tip of his finger in water and cool my tongue. I am in anguish in these flames.'

25 "But Abraham said to him, 'Son, remember that during your lifetime you had everything you wanted, and Lazarus had nothing. So now he is here being comforted, and you are in anguish. 26 And besides, there is a great chasm separating us. No one can cross over to you from here, and no one can cross over to us from there.'

There are four notable things in this lecture. First, that they were in an alert stage. Second, that it was not just the rich and Lazarus, since Abraham answers and says that there is a peak placed between you and us, making it clear that there were a lot of people. Third, that those that have died in Christ are resting and happy, while the non-saved are in torment. Fourth, even though they are not on the lake of fire yet, it seems as if the lake is near them because they can feel the flames and heat that torments them. I should clear up that even though Abraham and Lazarus were in that place, as well as David, Job, and all of the dead from the Old Testament, they are all saved. They are no longer in this place. Psalm 49:15 says, but God will redeem my life from the Sheol, because He will take me with Him. Like I mentioned, the dead from the Old Testament were there. Lazarus spent very little time there, because when Jesus dies and was buried for three days and three nights, what is mentioned in Psalm 49:15 was fulfilled. And He remembered Job; Job 14:13 says,

> *"I wish you would hide me in the grave[a]*
> *and forget me there until your anger has passed.*
> *But mark your calendar to think of me again!*

When Jesus resurrected, He brought with Him the saved that were in the Sheol, transporting them to paradise fulfilling the promise He made in John 12:32,

> *And when I am lifted up from the earth, I will draw everyone to myself."*

> *Ephesians 4:8-9, That is why the Scriptures say, "When he ascended to the heights, he led a crowd of captives and gave gifts to his people."*

> *9 Notice that it says, "he ascended." This clearly means that Christ also descended to our lowly world.*

This is why He said to the burglar that was next to Him on the cross in Luke 23:43,

And Jesus replied, "I assure you, today you will be with me in paradise."

2 Corinthians 12:2-4, I was caught up to the third heaven fourteen years ago. Whether I was in my body or out of my body, I don't know—only God knows. 3 Yes, only God knows whether I was in my body or outside my body. But I do know 4 that I was caught up to paradise and heard things so astounding that they cannot be expressed in words, things no human is allowed to tell.

In reference to that, Revelations 14:13 says:

And I heard a voice from heaven saying, "Write this down: Blessed are those who die in the Lord from now on. Yes, says the Spirit, they are blessed indeed, for they will rest from their hard work; for their good deeds follow them!"

In the same book of Revelations, we find another frame of those that were waiting in paradise for God's jury.

Revelations 6:9-11, When the Lamb broke the fifth seal, I saw under the altar the souls of all who had been martyred for the word of God and for being faithful in their testimony. 10 They shouted to the Lord and said, "O Sovereign Lord, holy and true, how long before you judge the people who belong to this world and avenge our blood for what they have done to us?" 11 Then a white robe was given to each of them. And they were told to rest a little longer until the full number of their brothers and sisters[a]—their fellow servants of Jesus who were to be martyred—had joined them.

Those that die in the Lord go to glory or to paradise, but those that are not saved go to hell until they are resurrected to face judgment, on the day of the grand white throne. It's sad to mention but, since the day they die, they go to hell and enter a great torment, as is the rich man mentioned in Luke 16:22-26 and those that have not arrived yet, until the final judgment, as per Revelations 20:12-15:

> *I saw the dead, both great and small, standing before God's throne. And the books were opened, including the Book of Life. And the dead were judged according to what they had done, as recorded in the books. 13 The sea gave up its dead, and death and the grave[a] gave up their dead. And all were judged according to their deeds. 14 Then death and the grave were thrown into the lake of fire. This lake of fire is the second death. 15 And anyone whose name was not found recorded in the Book of Life was thrown into the lake of fire.*

There are the people that Jesus refers to in Matthew 5:21-22 and 25:41 already mentioned. That makes it clear that hell is the waiting place until they are judged. And on that day, hell, and everyone in it, that have been condemned, will be tossed to the lake of fire.

There are cults like Growing in Grace *and Always Saved* that wrongly teach that once a person is saved, they are no longer able to be condemned or lost. But that is not what the Bible teaches.

> *Hebrews 10: 26-29, Dear friends, if we deliberately continue sinning after we have received knowledge of the truth, there is no longer any sacrifice that will cover these sins. 27 There is only the terrible expectation of God's judgment and the raging fire that will consume his enemies. 28 For anyone who refused to obey the law of Moses was put to death without mercy on the testimony of two or three witnesses. 29 Just think how much worse the punishment will be for those who have trampled on the Son*

of God, and have treated the blood of the covenant, which made us holy, as if it were common and unholy, and have insulted and disdained the Holy Spirit who brings God's mercy to us.

And verse 39 says:

But we are not like those who turn away from God to their own destruction. We are the faithful ones, whose souls will be saved.

The catholic church teaches that after death, they have to go through a purification process, known as purgatory, in order to be saved or be able to present themselves before God. This is a contradiction to 2 Corinthians 5:10:

For we must all stand before Christ to be judged. We will each receive whatever we deserve for the good or evil we have done in this earthly body.

Revelations 20:12, [12] I saw the dead, both great and small, standing before God's throne. And the books were opened, including the Book of Life. And the dead were judged according to what they had done, as recorded in the books.

Hebrews 9:27, And just as each person is destined to die once and after that comes judgment

If there were other means like purgatory, or always saved to reach salvation, Christ's sacrifice on the cross wouldn't have merit, and a life of sanctity like the divine laws demand, wouldn't be necessary. The bible verses we just read, wouldn't make sense either.

After the rapture and the great tribulation that will take place for seven years, there will be a period of judgments and events. The devil will be taken prisoner and tied in prison for a thousand years.

Revelations 20:1-2, Then I saw an angel coming down from heaven with the key to the bottomless pit[a] and a heavy chain in his hand. 2 He seized the dragon—that old serpent, who is the devil, Satan—and bound him in chains for a thousand years.

During these 1,000 years, Christ will govern the world and will rule it with iron rod, finishing the oppression and injustice.

Psalms 2:7-9, The king proclaims the Lord's decree:
"The Lord said to me, 'You are my son.
   Today I have become your Father.
8 Only ask, and I will give you the nations as your inheritance,
   the whole earth as your possession.
9 You will break them with an iron rod
   and smash them like clay pots.'"

Revelations 20:6, Blessed and holy are those who share in the first resurrection. For them the second death holds no power, but they will be priests of God and of Christ and will reign with him a thousand years.

After 1,000 years, the devil will be lose for a while. We don't know exactly for how long, but he will be arrested by the Angel and tossed to the lake of fire for eternity.

Revelations 20:3, The angel threw him into the bottomless pit, which he then shut and locked so Satan could not deceive the nations anymore until the thousand years were finished. Afterward he must be released for a little while.

Meanwhile, the saved during this time, the saved during the great tribulation, after the rapture of the church, the resurrection of the dead, the judgments, and all the processes of the events that will occur, we will be with the Lord. The Lord will be back with the church, and we will be with Him forever.

> *Revelations 6:12-14, I watched as the Lamb broke the sixth seal, and there was a great earthquake. The sun became as dark as black cloth, and the moon became as red as blood. 13 Then the stars of the sky fell to the earth like green figs falling from a tree shaken by a strong wind. 14 The sky was rolled up like a scroll, and all of the mountains and islands were moved from their places.*

Earth will go through this jury process, but God will make a new sky and a new earth.

> *Revelations 21:1-3, Then I saw a new heaven and a new earth, for the old heaven and the old earth had disappeared. And the sea was also gone. 2 And I saw the holy city, the new Jerusalem, coming down from God out of heaven like a bride beautifully dressed for her husband.*

> *3 I heard a loud shout from the throne, saying, "Look, God's home is now among his people! He will live with them, and they will be his people. God himself will be with them.*

> *Revelations 21:24, The nations will walk in its light, and the kings of the world will enter the city in all their glory.*

> *Revelations 22:5, And there will be no night there—no need for lamps or sun—for the Lord God will shine on them. And they will reign forever and ever.*

It leaves it clear that planet earth will continue to function after being transformed by God, and our Lord and the saved, we will be reigning here for all eternity. Take notes all the Christians that want to change the climate and think they can save the planet.

This information is not necessary in the order of how it will happen, it is just a list of things that will happen to prove His eternal existence. What has been mentioned is in the future; to be able to get to that future we need to live the present according to the word of God.

In reference to the present times, in the book of Matthew 24:12 says, sin will be rampant everywhere, and the love of many will grow cold. When we analyze the behavior of many Christians today, we can see we are currently living those times. We truly don't know who a true Christian today is. The word says by their fruits they will be known. Just because they go to the temple 2-3 times a week, and in some cases just on Sundays, week per week, or year by year, they are not delivering the fruits mentioned by God in the bible. To me it's like a tree that has been planted for many years but hasn't given fruits. What is it good for?

Many churches seem not to be in tone with the word of God. The shift with the way the world is living and do the things the world does. When I say a lot of churches, It's because we can't generalize because there a still many churches that live by the word of God. When I talk about churches, I am referring to congregations, because the church of God is one. I think of these last ones, there are very few.

> *Matthew 5:13-14, "You are the salt of the earth. But what good is salt if it has lost its flavor? Can you make it salty again? It will be thrown out and trampled underfoot as worthless.*

14 "You are the light of the world—like a city on a hilltop that cannot be hidden.

The church is the light of the world to dissipate the darkness of moral ignorance; and is the salt of the world to preserve the moral corruption. I have been in church since I was born, and I have seen how the church has changed. For example: the way people dress, the worship and how it's been modernized, and in all senses, it has been transformed.

Famous theologizes have taught us that since the first decade of the church, they conducted two types of services: one was of prayer, worship, and preaching; and the other was full worship known as party of love (Agape). To this party, only believers were able to assist. According to historians, in the first century spiritual songs were being written and sung with Psalms. It wasn't until like 30-40 years back, that all types of music were introduced.

An example of this is the following.

I was leading worship at a church for more than a year. After a year, the church chose me as Worship Leader in the annual assembly where the people that will be working are chosen. The Pastor and the officials gave me my responsibilities and rights in writing, as a minister of that department. In the rights and responsibilities, it was established that I was fully responsible for the music and the worship ministry at the church. A couple of months later, on two occasions, I arrived ready to execute the worship plan I organized on Saturday, and I found that two of the officials had brought musicians from outside to minister and lead worship, without even letting me know, violating the same contract they had given me. I had a meeting with the pastor and the officials to see how this situation could be resolved, since it was unfair to prepare a program, do rehearsals and walk into this situation. The following were the words spoken by one of the elders of the church (verbatim): "Listen Brother Del Toro, without a doubt, you sing and touch very good the guitar, but your songs are outdated and are very conservative, but we need you to minister with more contemporary music." This information was surprising to me especially coming from a deacon. I believe that when we go to church it is with two main purposes. There are many reasons, but these two are essential.

(1) Worship God through singing, prayers, and offerings. Even the Webster dictionary states that to worship is to revere, respect, and give honor to the supernatural power. That is to the Supreme. I believe that in the manner that most people behave, since they set foot in the temple, says a lot, in regard to the respect and honor to God. One credit that I give the Catholic church is that since the moment people walk in, to the time they leave, they do it with a lot of reverence.

(2) For God to talk to us through the preaching or through the gifts of the spirit.

In the majority of the cases, no matter how spiritual the people may be, they don't arrive to service completely prepared to offer worship for various reasons. I believe the person that is in charge of the worship is required to prepare the environment with chants and prayers. The first songs should be happy toned so that people can leave behind their worries and can step into worship. Then, after the atmosphere is ready, step into a profound intimacy with the Lord with those songs that the Bible tells us. There are people that think a devotional dedicated to God is something practical or theorical. On one occasion a pastor told me that in the church he pastored, they had the practice of singing a fast toned song first, and then a slower one. Every time I lead any service, my only purpose was to prepare a spiritual atmosphere, and not allow that nothing interrupt that communion with God, until I handed the rest of the service to the pastor. I believe that when a person or many people are dedicating a service to God, wither it's in a church or home, they should do it with a lot of reverence, since we are communicating with the Almighty. I have seen people give honor to other men like the president or other deputies. Imagine how much more if we are talking to God. If anyone needs to know what type of respect I am referring to, read the requirements that any person has to have when trying to visit the Queen Isabel. That is for Queen Isabel, imagine how much more when it is regarding the King of Kings, our Almighty, who deserves all glory. The suggestion from this deacon was clear to me. These types of Christians are the ones that have converted churches into entertainment clubs. The bible teaches us in the book of Colossians 3:16:

> *Let the message about Christ, in all its richness, fill your lives. Teach and counsel each other with all the wisdom he gives. Sing psalms and hymns and spiritual songs to God with thankful hearts.*

For example, Isaiah 44:22, Isaiah 12:6; Psalms, and many other inspirations of men and women of God. Ephesians 5:18-19:

> *Don't be drunk with wine because that will ruin your life. Instead, be filled with the Holy Spirit, 19 singing psalms and hymns and spiritual songs among yourselves and making music to the Lord in your hearts.*

This is the type of worship that needs to be given.

The excuse that many people use is that music was created by God, and that is true. Music is creation of God. What they don't say is that, just like many other things, like sex for example, is also God's creation. But the devil has used it as instrument or a means of perdition. Satan was a creation of God, as a light cherub, and rebelled against Him, wanting to be like Him. Isaiah 14:12-19:

> *"How you are fallen from heaven,*
>> *O shining star, son of the morning!*
> *You have been thrown down to the earth,*
>> *you who destroyed the nations of the world.*
> *13 For you said to yourself,*
>> *'I will ascend to heaven and set my throne above God's stars.*
> *I will preside on the mountain of the gods*
>> *far away in the north.[a]*
> *14 I will climb to the highest heavens*
>> *and be like the Most High.'*
> *15 Instead, you will be brought down to the place of the dead,*
>> *down to its lowest depths.*
> *16 Everyone there will stare at you and ask,*
> *'Can this be the one who shook the earth*
>> *and made the kingdoms of the world tremble?*
> *17 Is this the one who destroyed the world*
>> *and made it into a wasteland?*
> *Is this the king who demolished the world's greatest cities*
>> *and had no mercy on his prisoners?'*
>
> *18 "The kings of the nation's lie in stately glory,*
>> *each in his own tomb,*
> *19 but you will be thrown out of your grave*
>> *like a worthless branch.*
> *Like a corpse trampled underfoot,*
>> *you will be dumped into a mass grave*

with those killed in battle.
You will descend to the pit.

Ezekiel 28:14, I ordained and anointed you
*   as the mighty angelic guardian.[a]*
You had access to the holy mountain of God
*   and walked among the stones of fire.*

I have sustained conversations with people that have told me they have visited churches and that they would not visit them again, because to listen to salsa and reggaeton, they can do that at home. They are people that are not Christian but have a true concept of what a church should be. Just imagine, if the elders of the church are the first ones called to give God their primates to the real worship, are the first to sink under the desires of the flesh; what can you expect from the ordinary members? It seems they haven't learned yet what Galatians 5:16-17 says:

> [16] *So I say, let the Holy Spirit guide your lives. Then you won't be doing what your sinful nature craves. 17 The sinful nature wants to do evil, which is just the opposite of what the Spirit wants. And the Spirit gives us desires that are the opposite of what the sinful nature desires. These two forces are constantly fighting each other, so you are not free to carry out your good intentions.*

These types of leaders are the ones responsible that the church of God doesn't grow. This is the same church that before I arrived to help them where a very big church in number and now were reduced to just a small group and a couple of visitors. When the church operates in base of mechanism and modernism, and is led by carnal people, no matter the efforts of the pastor, the church will not grow. The church will only grow when it is led by the Holy Ghost, and not based on beliefs, culture, or personal tastes. In many churches, these types of officials don't even allow the pastor to do their jobs as they wish. Their jobs should be to help the pastors since they are the spiritual leaders; the man God has placed to lead and direct the church.

In many churches, the officials feel they are the owners of the church. These are the officials that don't even pray one hour a week, and don't fast once a year. They are like the church of Ephesians in the book of Revelations, chapter 2, that they lost their first love and only serve to be recognized by the works of the flesh and they don't allow the Holy Spirit to work and be Him who leads the church.

During the many years that I have been serving the Lord, I have seen how many churches have divided, and pastors have had to resign and leave, due to these types of leaders. These types of people should never be named leaders of any church since they don't have the requirements that are established in the book of Timothy, Titus, and other books of the Bible. The coldness that many Christians show reflects the lack of true spiritual leaders. It is worrisome, because although there are a few left, there are less and less each day.

Spiritual leaders like Paula Lorenza Lugo (Loren), my mother, prayed for 3 to 4 hours every day, fasted weekly, and was a wise woman in all aspects. Despite having graduated with honors from the Universidad Interamerican de Puerto Rico (a private college in Puerto Rico), and be a real intelligent woman, she was a humble woman. She was full of God's power. The evidence was in the fact that every time she was called to intervene with a person that was full of demons, when she rebuked them, they would immediately leave. I remember there were moments she was called to a home, before she opened the door to go in, the demons that possessed the persons would scream and say, why did you bring this woman here?

Now we have to go! I will never forget when I was about 10-11 years old, there was a man that lived in the neighborhood and he was possessed by demons, and they would lock him up nailing the doors and windows. He would break them and start running around terrorizing people. One Sunday he entered the church and started screaming. Four brothers in the faith tried to control him but couldn't pin him down with all of his supernatural strength. My mother was praying at the altar like she used to during the devotional. She got up when she heard the screaming. She walked over to him; his name was Victor. She told the men that were trying to control him to let him go. When they let him go, Victor stayed still. She called out for someone to bring a chair. She ordered him to sit. He sat down looking down to the floor. My mom told him to look at her and he responded that he couldn't. My mom immediately told the demons that were controlling him to let him go. Immediately he became free. He lived many more years after this and never again was possessed by demons. She was a tremendous leader, with a lot of wisdom and would resolve any issues for the greater good and full of peace. She was a true teacher. Not just in the natural world, but spiritually as well. God used her in divine healing and helping the people in the community even if they weren't believers.

To write about this true spiritual leader, and all of her powerful testimonies, and works she did as a leader, from the battles she sustained with the devil, and her experience to have died and resurrected after 3 brain strokes, it would take various books. In one of her testimonies, having her right side completely dead, and not being able to pronounce half a word because of the brain strokes and stranded to the bed for seven months, not being able to sit, and forsaken by the doctors. In an instant miracle, God lifted her, and her body started to function normally. She started to walk and talk. This was a total miracle acknowledged by doctors! I want it to be clear that she never went to physical therapy. In her condition, not even that was able to be coordinated. Months later, the same doctors that sent her home to die, when they saw her walking in the hallways of the hospital, they almost fell on their backs. They were so surprised because they thought that months ago, she should have been buried. Because of this condition, before she was sent home, she died twice. The first death, certified by doctors, was for three hours. She was already in the morgue, waiting for the funeral home to pick her up. One employee observed that the white blanket that was covering her was moving, and he ran to advise that she was alive. She was immediately moved to a recovery room. Five doctors were working on her, including the ones that had declared her dead. While they were working on her they said they didn't understand what was happening. She died a second time, on this occasion for half an hour. But God brought her back to life. After this she worked for another 25 years like she had always done in the works of the Lord. Generally, when God has spiritual leaders of this level, they are people attached by Satan. For example, the testimony of the brother Yiye Avila. Satan tried to eliminate his ministry, when his son-in-law killed his daughter, stabbing her multiple times. What this man of God did was arrive at the prison where the man that killed his daughter was, he forgave him and told him that God loved him. You can't find these types of spiritual leaders anymore.

The following are requirements for elders and bishops in the works of the Lord. Titus 1:7-9

> *A church leader[a] is a manager of God's household, so he must live a blameless life. He must not be arrogant or quick-tempered; he must not be a heavy drinker,[b] violent, or dishonest with money.*

⁸ Rather, he must enjoy having guests in his home, and he must love what is good. He must live wisely and be just. He must live a devout and disciplined life. ⁹ He must have a strong belief in the trustworthy message he was taught; then he will be able to encourage others with wholesome teaching and show those who oppose it where they are wrong.

On verse 13 of the same chapter, it says:

This is true. So, reprimand them sternly to make them strong in the faith.

Revelations 7 says: anyone with ears, must listen to the Spirit and understand what He is saying to the church.

Regarding the clothes, I am not referring to a dress or pants, but to dress like the word of God tells us in 1 Timothy 2:9, to dress honestly and with shame. Today there are women that walk into the church with provocative clothing, occasionally showing off everything, and in many churches, this is normal. It's not that we are old fashioned, or that the half to wear dresses down to their ankles, but rather they dress like Christian women, in accordance with the word. For example, if a woman dresses with loose pants and another with a short, tight dress, showing almost everything, who is better dressed? It is sad to say but the majority of Christian churches have turned into social clubs. This is the reason why the gifts of the Spirit don't operate in the majority of the churches. The proof is that years ago when you walked in the church, immediately you would see the gifts of the Spirit work in tongues, discernment, science, interpretation of the tongues, divine sanity, prophecy, and a genuine movement of the power of God. Where is all of that today? And why? The Spirit of God doesn't operate based on emotions and joy of the flesh. I don't want to say with this that we can't get joyful or get excited since that is our human nature, but in the spiritual, we have to do it as the word of God says. God wants worshipers that worship in spirit and truth. John 4:23.

I don't know if you have noticed, but in church when they are singing a song that is very moving, almost to a point where you can dance (but dance in the spirit), people start shaking and speaking tongues. But then when the music stops the tongues disappear and all emotions stop. I write it like this because for me they are just emotions, human or carnal tongues, and joy in the flesh. I believe in the movement of the Holy Ghost, but in a genuine movement that with or without music, you could still feel the presence of the Holy Spirit, working and healing, not just in the physical but also in the soul; convicting so we can be full of truth, and we end up edified, healed, restored, and blessed. All of these behaviors and the ones I am going to mention, are the cause for many people to have lost their faith and the interest to visit a church. I guarantee that the churches that are still light in the world and allow the Holy Spirit to guide them, people come and go truly blessed, healed both physically and spiritually. They are churches that get filled and it's a pleasure to visit them. It provokes their members to invite people to their services like Trinity International Church in Lake Worth, Florida. It has more than 7,000 members and is guided by a true man of God, Pastor Tom Peters. This is also the case of Victory Church in Lakeland, Florida, directed by the man of God and Pastor, Wayne M. Blackburn. It is not a coincidence that it also has thousands of members. I say it's not a coincidence because when the gifts of the spirit, when the Pastor and leaders are subjected to God, and operate in the church, the results are positive. We know that many years ago there were dogmas and beliefs that were not according to the word of God, but without doubt, the church has gone from one extreme to the other. There are Christians that you can't mention these things to because they will get defensive on their postures and say that we are old fashioned, and that we haven't realized that times have changed. Like I have mentioned, Hebrews 13:8, Jesus Christ is the same yesterday, today, and forever. He doesn't have a variation shadow. In some things that have nothing to do with the spiritual aspect, they are right, we have to be up to date. For example, the computers and internet programs, are good progress changes for the church. But in the spiritual aspects, no. We need to have our minds like Christ as the word says in John 3:11,

> *I assure you; we tell you what we know and have seen, and yet you won't believe our testimony.*

In Hosea 4:6 the bible says: *my people perished because they didn't have knowledge.* In the book of Matthew 24:12 the word says: *sin will be rampant everywhere, and the love of many will grow cold.* Verse 22 says: *In fact, unless that time of calamity is shortened, not a single person will survive. But it will be shortened for the sake of God's chosen ones.* The days we are living it is very notorious to see a large quantity of Christians, have their sight on things that perish. The bible teaches us in Hebrews 12:2, *we keep our eyes on Jesus, the champion who initiates and perfects our faith.*

WHY AS CHRISTIANS WE SHOULDN'T SUPPORT POLITICIANS THAT HAVE DECLARED TO BE ENEMIES OF GOD

Depending on the political views of the reader, the content of this information can be sweet or sour. What is guaranteed is that what I have written here is a complete reality, and not a reality on my point of view, but in all of its content, what history and the records show as well as the coldness and behavior of Christians today. This is a reality that for many that have their customs and traditions well rooted, it will be hard for them to swallow, both in the spiritual and in the political. I mention in the political, because the same way I will mention it later on, of how a credit card can affect our spiritual life, the same way we can become aware, the political decisions can also harm our spiritual life. And even worse, it is already affecting the works in the church. The politicians have been the ones to maintain the separation between church and state, since this has been the smart works of Satan, to keep the church silent. I also want to make clear that the rest of the information I share it is not defending any political party, but rather defending what God has established. Unfortunately, there are political parties whose ideologies are against what has been established by God. Everything that is opposed or that legislates against what God has established, will never have my support, and shouldn't have the support of any Christian that professes to love God above all things. I say love God above all things because what I have seen at the time to go vote, is that Christians are more interested in the politician that promises them social benefits, even if they are against what God has established, they give them their vote. The information should be analyzed with all honesty and transparency if we want to see and understand the realities that as Christians should interest us, in the biblical context and not in the political. There are a lot of things happening today, not just in Washington D.C., but all over the world, and especially in the middle east, and it has a lot to do with the fulfillment of the word of God. There are many alleged Christians that I have no idea what knowledge of the word of God they have because the boast saying that they are liberals, and that they agree with the political postures of liberal politicians. For me as a Christian, how sad and painful that sounds, since that is the same as being in favor of abortions, same sex marriage, and the long list of the things the liberals approve and disapprove.

Of the prophetic knowledge of the sacred scriptures, the politicians today don't even know where they stand. It is very noticeable the difference between politicians today and politicians from years ago. Years ago, all the meetings in the congress started with a prayer. When you read the constitution of this country, you can tell it was done based on the Bible. The politicians from the past had God very present, even in their conversations and the laws the legislated. A clear example is the letter that President George Washington on the 3rd of October 1789, assigned the third day of November, every year, as the day all of the nation needed to thank God for his great provisions, his favors, his mercy, his protection, the peace, the union, and for having allowed to write the constitution of this nation for its security and happiness. These are the words quoted that are found in this ordinance. Anyone that has biblical knowledge, when they read this ordinance completely, they will be able to identify the vocabulary and the biblical knowledge in which this president expresses himself. We should also mention that this was a petition that Congress made to President Washington in 1787, so that this declaration be done. Democrats today, in vast majority, have no regards for God. Quite opposite, they have an arrogance of power and believe to be over the Creator. This is why you see that they take away rights to Christians of praying and talking about God in school and other public places, but they grant all the rights to homosexuals to express and behave freely, however they want. That is why I can't understand how Christians support these candidates. For example, in all of the history of Congress, the committees were started acknowledging God and swearing the witnesses before God. I was watching a live session directed by the Democrat Jerry Nadler, and they eliminated the acknowledgement and recognition before God that I mentioned. One of the Republicans protested the elimination of what had been established since the beginning of Congress, and Mr. J. Nadler answered that congress wasn't a religious institution. As Christians we need to examine the acts of the politicians and not just what they talk or write. For example, on many occasions, especially in presidential debates, for political convenience, seeking the Christian votes, they state to be Christians, religious, and people of faith. Nevertheless, their acts say the contraire. In the debate, they allege to be Christians, and in the same debate when they are asked if the favor abortions, now up to nine months, they say they agree. The say they are people of faith, but aware they lie and defame and do whatever possible to defeat their political enemies. The lack of knowledge like I have mentioned before, bring bad results. Many times, we make mistakes because we don't have the knowledge, and we don't have it because of apathy and lack of interest. Other times, even though we have the information, we don't think, and we don't analyze the results of our decisions. We can use as example a person that has a credit card. They know that getting into debt will bring them problems, but at the moment they use it, they don't analyze, and the results can be catastrophic. Imagine if its catastrophic, that they come to church and are not able to pray or sing with liberty and joy, because they are so worried about their economic situation. This teaches us that even things we think are small and insignificant, jeopardize our spiritual life, our joy, and our internal peace.

The knowledge that the word talks about, is spiritual knowledge. Colossians 1:9 says *God fills you with knowledge and spiritual wisdom.* A Christian that has spiritual knowledge, doesn't get tangled in the rudiments of the world and doesn't put his convictions and beliefs, over the spiritual ordinances.

The following illustration is an example of how the lack of spiritual wisdom, we suffer bad consequences.

There was a Christian man that lived on the shore of a river. One day, after strong rains, the river overflowed, and the house began to flood. Someone in a small boat passed and offered him to come on board so he wouldn't perish. He said no because he trusted that God wouldn't allow him to perish. The rising grew more, and the man climbed on the roof of his house. A helicopter came and he refused to be saved because he trusted that God was going to save him. The man drowned and went to heaven. As soon as he came before God, he asked him: Why did you let me drown? I had all my faith in you, and you didn't save me. God answered: I sent you a small boat, then a helicopter and you didn't let them save you. This humorous tale has a lot of truth in it, and it is as simple as the result of ignorance. In old times, there were Christians that refused to drink medications, and in some cases not even an aspirin. They honestly thought that they would offend God for not having faith. Only God know how much suffering they would have avoided. Medicines are a means that God uses and it's precisely God who allows them to be found, invented, and created. God himself in Jeremiah 33:6 says:

> *Behold, I will bring it health and cure, and I will cure them, and will reveal unto them the abundance of truth and peace.*

> *Ezekiel 47:12, Fruit trees of all kinds will grow on both banks of the river. Their leaves will not wither, nor will their fruit fail. Every month they will bear fruit because the water from the Sanctuary flows to them. Their fruit will serve for food and their leaves for healing.*

The Bible is more than a university, it contains all types of knowledge, not only spiritual, but even the simple details of life, all the way to the more profound, from the beginning of the world, until eternity. If they scrutinize the Bible which is the word of God, they have enough knowledge the make the correct decisions while we are here on earth, and not be deceived by the press, and corrupt politicians that are God's enemies. I write this based on what many Christians share in social media that confirms the lack of spiritual knowledge that they have.

The following has to do with our political decisions, that also have to do with our spiritual life. Having the knowledge I have of the word of God, (and it's not that I know it all), but rather I have the sufficient theological wisdom to discern what is and isn't acceptable to God. I've been gaining knowledge for many years from the ideologies and beliefs, together with the history of the Washington

Legislators. That is why it is so surprising to me, not only the number of Christians that vote for Democratic candidates, but how they defend their ideologies and beliefs. Based on the word of God, how can a Christian support a Democrat that say they don't have an issue on a woman having an abortion of a baby, even if it's about to be born? (Because Hillary Clinton said, if they are not born yet, they don't have constitutional rights). I don't know what to think of a person with that mentality. Even less, what to think about a Christian that supports an enemy of God. We have come to a point that there are politicians like the ex-president Obama and others from the Democratic Party, that agree that if they do an abortion and the baby is born alive, it should be tossed aside and let it die, since that was the plan.

The reasons Christians support these types of candidates are traditions, culture, education, influences, and convenience.

The following information is exactly what both parties, Democrats and Republicans, support, and claim.

All of this information can be verified in the Congress records, as well as different internet sites such as: https://www.senate.gov/.

Also, in the history books that have to do with laws and the legislators of this country; those that have presented the laws, as well as those who have voted in favor or against them; and not by what the press informs that intentionally lie, inform based on speculations, omit certain truths, and in other cases they twist the truth to accommodate their ideological postures. This is just an example of many: The Republicans and some Democrats oppose to abortions for two reasons: the first being that they are prolife. Better said, they are not in agreement for those lives to be killed. The second being the main reason why they are opposed: they don't want the billions of dollars from the taxpayers to be used to pay those abortions every year. How has the press twisted the information? What they report is: (1) that the Republicans want to dictate to women what they can or can't do with their wombs; (2) that they want to prohibit their rights. The truth is that they believe it is unmoral and criminal, as well as contraire to what God has established. Exodus 20:13, Thy shall not kill. 1 Samuel, Jehovah kills, and he gives life. He makes descend to Sheol and makes it rise. Acts 17:24-25:

> ***The God who made the world and everything in it is the Lord of heaven and earth and does not live in temples built by human hands, as if he needed anything. Rather, He Himself gives everyone life and breath and everything else.***

And like I mentioned that the tax money that we pay should not be used for this purpose. I believe it's clear that even before it was approved, there have always been abortions. But they were abortions done with the government's permission. Now, thanks to the Democrats, they have permission or license to practice these abortions, that according to the reports, since these abortions were approved, there have been over 50 million abortions executed. Based on what has been informed, it is not my interest, or the interest of the prolife legislators to tell any woman what they can or can't do with their womb.

WHY AS CHRISTIANS WE SHOULDN'T SUPPORT POLITICIANS THAT HAVE DECLARED TO BE ENEMIES OF GOD

THE POLITICAL PLATFORM OF BOTH PARTIES

The following, without adding or subtracting, is the platform, and what both parties, Democrats and Republicans, approve and disapprove.

POSITIONS OF REPUBLICANS AND POSITION OF DEMOCRATS

TRADITIONAL MARIAGGE IN FEDERAL RIGHT

Support the federal law of matrimony defense (DOMA)

REP (Yes) DEM (No)

CLONING

Support the human cloning

REP (No) DEM (Yes)

OPPOSE TO JUDICIAL ACTIVISM

REP (Yes) DEM (No)

ENERGY

Amplified perforation for petroleum

REP (Yes) DEM (No)

HUMAN LIFE

Support the protection of a child's life that is born alive and survives a failed abortion

REP (Yes) DEM (No)

HOMOSEXUAL EDUCATION

Supports the education plan that promotes homosexuality

REP (No) DEM (Yes)

BUSINESS FREEDOM

They are against the laws that force businesses to favor the homosexualism

REP (Yes) DEM (No)

ARE AGAINST GAY PRIDE AND GAY MARIAGGE

REP (Yes) They refused to support the celebration of Gay Pride and Gay Marriage

DEM (No) They support the Gay Pride and Gay Marriage

YOUTH AND ABORTION

Support the transport of minor girls through the state boarders in search of a secret

Abortion without the parents concernment.

REP (Yes) DEM (No)

WEAPONS RIGHTS

They are against a prohibition of an assault weapon

REP (Yes) DEM (No)

PARTIAL BIRTH ABORTIONS

They are against abortions through partial births

REP (Yes) DEM (No)

TRADITIONAL MARRIAGE IN STATES

Support the state amendments of marriage

REP (Yes) DEM (No)

PARETAL RIGHTS IN EDUCATION

Supports the parents election of schools in the education

REP (Yes) DEM (No)

SUPPORT GAY MARIAGGE

REP (No) DEM (Yes)

All of these questions and answers should be analyzed through the word of God, and not through the political interests of corrupted people, that don't have the knowledge of the word of God, and even less fear Him. Their only purpose is, satisfy their own ego and make money.

It is more than clear that the Democratic platform is the one that approves everything that goes against the word of God. When I talk about the political platform, it's because this is what this party mainly approves. More or less, it's their ideology. And we know, ideology is a group of ideas and form of thinking that characterizes a person. There is a minority of Democratic Conservatives that do not agree, and a minority of Republicans that are liberals and do agree. In my opinion, these types of Republicans, shouldn't receive our votes since they are acting as the liberals, legislating against the word of God. I want to make clear that these groups of minorities are mainly at state level, because in the Senate and in the National Congress, they don't exist. I also don't have an issue to give the conservative Democrats my vote that oppose to legislate against of what has been established by God. The only problem with that today is that is very difficult, at least at Congress and Senate level, find at least one Democrat Conservative. Years ago, you were able to find them. I believe the last conservative president was Kennedy; but the Democrats today compared to Democrats 40 years ago, they have done a 180 degree shift, and have become enemies of God and the church. What I have seen in many Christians is that they are more interested in the social benefits than the ordinances God assigned.

Like I mentioned before, many people are lying to receive the social benefits already stated; and there are many Christians that are going down the same path and doing the same thing. We have to scrutinize the word of God teaches us in reference to the mentioned above.

> *2 Thessalonians 3:6-12, In the name of the Lord Jesus Christ, we command you, brothers and sisters, to keep away from every believer who is idle and disruptive and does not live according to the teaching you received from us. For you yourselves know how you ought to follow our example. We were not idle when we were with you, nor did we eat anyone's food without paying for it. On the contrary, we worked night and day, laboring and toiling so that we wouldn't be a burden to any of you. We did this, not because we do not have the right to such help, but in order to offer ourselves as a model for you to imitate. For even when we were with you, we gave you this rule: "the one that is not willing to work shall not eat." We hear that some among you are idle and disruptive. They are not busy; they are busybodies. Such people we command and urge in the Lord Jesus Christ to settle down and earn the food they eat.*

DISOBEDIENCE WILL BE CATASTROPHIC

When the Lord comes to rise His church, there will be a lot of surprises. Many Christians believe that living and behaving like the mundane, they will be lifted. This is why God Himself warns us. Many are called, but few are chosen.

Revelations 13:8 says: [8] And all that dwell upon the earth shall worship him, whose names are not written in the book of life of the Lamb slain from the foundation of the world, what did they mean with this? That God in his omniscience already knows before we are born, who will be obedient to his ordinances, and who won't be. Who are the disobedient that play religious roles, but their hearts are far from the Lord. A list of some things starting with idolatry. 1 Corinthians 6:9-10 [8] And all that dwell upon the earth shall worship him, whose names are not written in the book of life of the Lamb slain from the foundation of the world.

Definition of idolatry, since some people think that kneeling before images is the only form of idolatry.1) Religious practice in which cult is rendered to an idol. 2) Excessive love and admiration that is felt for a person or thing. For example, the Pope, Queen Isabel, Obama, etc.

Deuteronomy 4:15-18, [15] *"But be very careful! You did not see the Lord's form on the day he spoke to you from the heart of the fire at Mount Sinai.* [16] *So do not corrupt yourselves by making an idol in any form—whether of a man or a woman,* [17] *an animal on the ground, a bird in the sky,* [18] *a small animal that scurries along the ground, or a fish in the deepest sea.*

Colossians 3: 1-17, Since you have been raised to new life with Christ, set your sights on the realities of heaven, where Christ sits in the place of honor at God's right hand. [2] *Think about the things of heaven, not the things of earth.* [3] *For you died to this life, and your real life is hidden with Christ in God.* [4] *And when Christ, who is your[a] life, is revealed to the whole world, you will share in all his glory.*

[5] *So put to death the sinful, earthly things lurking within you. Have nothing to do with sexual immorality, impurity, lust, and evil desires. Don't be greedy, for a greedy person is an idolater, worshiping the things of this world.* [6] *Because of these sins, the anger of God is coming.[b]* [7] *You used to do these things when your life was still part of this world.* [8] *But now is the time to get rid of anger, rage, malicious behavior, slander, and dirty language.* [9] *Don't lie to each other, for you have stripped off your old sinful nature and all its wicked deeds.* [10] *Put on your new nature and be renewed as you learn to know your Creator and become like him.* [11] *In this new life, it doesn't matter if you are a Jew or a Gentile,[c] circumcised or uncircumcised, barbaric, uncivilized,[d] slave, or free. Christ is all that matters, and he lives in all of us.*

[12] *Since God chose you to be the holy people he loves, you must clothe yourselves with tenderhearted mercy, kindness, humility, gentleness, and patience.* [13] *Make allowance for each other's*

faults and forgive anyone who offends you. Remember, the Lord forgave you, so you must forgive others. [14] *Above all, clothe yourselves with love, which binds us all together in perfect harmony.* [15] *And let the peace that comes from Christ rule in your hearts. For as members of one body, you are called to live in peace. And always be thankful.*

[16] *Let the message about Christ, in all its richness, fill your lives. Teach and counsel each other with all the wisdom he gives. Sing psalms and hymns and spiritual songs to God with thankful hearts.* [17] *And whatever you do or say, do it as a representative of the Lord Jesus, giving thanks through him to God the Father.*

All the disobedient that practice sin, and don't want to listen, since hearing is one thing and listening is another, are those whose names are not written in the book in life. The reason for the disobedient is found in Philippians 3:18-19

> [18] *For I have told you often before, and I say it again with tears in my eyes, that there are many whose conduct shows they are really enemies of the cross of Christ.* [19] *They are headed for destruction. Their god is their appetite, they brag about shameful things, and they think only about this life here on earth.*

One thing is to commit a sin, or an involuntary sin, and another is to voluntarily sin, or to practice sin. This is clear in John 1:8:

> *John himself was not the light; he was simply a witness to talk about the light.*

> *1 Juan 3: 8-9,* [8] *But when people keep on sinning, it shows that they belong to the devil, who has been sinning since the beginning. But the Son of God came to destroy the works of the devil.* [9] *Those who have been born into God's*

family do not make a practice of sinning, because God's life[a] is in them. So, they can't keep on sinning because they are children of God.

We should examine word for word, what has been said, to see which of the ordinances we disobey and practice. For example, the pontiff of the Catholic church in 2008 expressed that he was upset with Catholics of this country for having chosen to vote for Mr. Barak Obama, having already shared his recommendation with them not to so, because of his position on abortion. I know a lot of Christians whose pastors advised them not to vote for Obama for the same positions of the Pope, they told me they would vote for Obama, because the country needed a change. A Christian made the following declaration to me: We have to vote for a change, and added, I won't vote for Mr. McCain for any reason. Mr. McCain chose a candidate that had a romance with another person. My response was the following. No press has certified that to be true. The alleged information was published by the newspaper "The Enquirer" an everyone knows the type of information they put out. This was just with the intention to cause defamation with the purpose to cause political harm and assure the Democratic Party a victory. I advised him on the danger of believing this type of information, because like him, there are many others that don't have clear the information. Mrs. Sarah Palin is a Christian woman, that assists at Assembly of God Church, and was ones that was accused by President Obama's campaign, by a group of 30 people that went to Alaska to try and find out as much dirt on her as possible. The report they gave was that she was an evangelic woman that went to a church that spoke in tongues. I remember the day they gave the news because the person that gave the news started to imitate the tongs, making fun of Christians. The same way there have been people that have had the courage and the nerve to make fun of her son with down syndrome, because she doesn't believe in abortion. They reported this because they don't think that is acceptable. For the liberal press evangelicals are crazy ignorant. This church came out in the news on December 15th because precisely the political opposes burned it down. Although it is true that some people in the Republican party did not support them because they consider it an extreme right, but for those that know, we understand what this means. In my opinion, we need more politicians like Sarah Palin., so we can come out of the shamelessness that we are living in and walk back to grace and the principles this nation was based on. Not because I heard on the Mr. Snichttsh's Show that Mrs. Sarah Palin was evangelic I believed she was a true woman of God, but rather, I heard her in the church, the way she spoke and the way she prayed, Because I know who can try to pass as Christian, I have no doubt that this is a woman of God.

Because of everything that I have expressed, I have no doubts of the lack of knowledge that many Christians have. How would you explain the political decisions of the votes towards the people that are against all the things God has established, and they go against the people that are in our favor; better said, against the church of Christ. This is an example. The apparent Christians never questioned President Obama's postures, aside from knowing that he is in favor of abortions and having voted in favor in his home state of Illinois, to terminate the lives of a baby that survive a botched abortion, he also has a record in Congress of being the Senator most liberal in all of the US Congress. He is not only in favor of abortions but of same sex marriage, and as well of many things that go against the word of God. All of these things appear in the registry for Mr. Obama.

The following information is evidence that Christians are following traditions and the information provided from the liberal press in reference to political decisions. In all honesty, who knew Obama before he was elected President? The true answer is no one. His resume doesn't contain anything that justifies his presidency of this nation. Mrs. Palin wasn't known either, but had a better political resume than Obama because, besides having several important political positions in Alaska, she was also the governor of this state with 75% approval of the population. The fact that the liberal press only attack Mrs. Palin stating that she wasn't capable to assume the presidency is another evidence that they manipulate the information to favor those that are allied to them. In regard to what has to matter to us as Christians, Mrs. Palin's resume was the one that mostly complied with the requirements. Aside that her candidature wasn't for the presidency, but because of her conservative postures, she was attacked by the press and many Christians. Knowing the word of God and the ideology of the Democratic Party, being the majority in both cameras and senate, it is too late for the decisions that this president will make that are against the word of God. That has been the posture of the Democrats for the last 50 years. The more liberal they are, the more harm they will cause. I am sure that this president will continue making decisions against what God has established and many Christians that voted for him will continue to walk like if nothing has occurred. Spiritually speaking I see it as walking themselves to be sheep for sacrifice. Records are clear demonstrating that the Democrats are responsible for removing the bible and prayers from schools. They are the ones that approve not being able to even speak about God in school hallways The Democrats are the ones that approve sexual education in schools and handing out birth control so that young girls avoid getting pregnant.

They also approve that if they get pregnant, they can get an abortion with or without their parents permission.

One of the worse things that also proves everything that has been informed here is that since abortions have been approved, there has been a war against Democrats and Republicans, that gets worse every day in Washington, because the Republicans are against the abortions getting paid with the taxpayer's money. As a Christian, this is unacceptable that we have to pay with our taxes for the things we are against thanks to the Democrats. They have disguised this sin under the name of Family Planning. This is how they are able to manipulate the population that refuses to get educated in this manner and only follow traditions. It is sad that there are many Christians tied down to support this type of legislators. Here in the United States, same sex marriage has been approved by the legislators, the signature of Democrat governors, and the approval of the Supreme Court. The ideology of both parties doesn't change, it will always be the Democrats that approve the decisions that go against what God has established, converting the country into another Sodom and Gomorrah. The Christians that support these candidates are directly and indirectly participating of these crimes and sins. Regarding abortions I call it a crime because once the heart begins to beat, it is a living creature, and it is a sin to kill them. I can't understand how easy it is for them to kill so many innocent babies and how many Christians are ok with this and later walk into church to sing Christ is coming soon, I am leaving with Him. Placing liberals in Congress and at the White House is a great threat for liberal judges to be nominated for these positions, that the news report, there will soon be 2-3 vacancies. If that is the case, they will become majority and it will be completely against the Church and the conservative Christians. We have been hearing many preachers warning that there will come persecution against the church, since this is what is said by the word of God.

THE PERSECUTION AGAINST THE CHURCH

A great Baptist Pastor and historian explained the following: since the year 1947 the liberals in this country have been fighting to terminate all Christian principle.

In 1962, the senator and atheist from the state of Texas Madelyn Murray O'Hair, recollected thousands of signatures and later went to the court demanding that the bible and prayers be taken out from public schools. The court composed of liberal judges in their majority, voted 6 to 1 in favor of this petition.

In another demand in January of 2005, by Mrs. Kay Stanley, a lawyer that also worked in Real Estate, demanded in Houston, Texas that the bible be removed from the courthouse in the Harris County. The court composed by a liberal judge, approved her petition, and ordered the bible to be removed. In an appeal by local officials to this demand, the circuit court of the United States, in New Orleans, Louisiana, with the majority of liberal judges, on August 25th of the same year, sustained the decision with votes 8-1 in favor. And as we all know, we no longer swear in by placing our right hand on the bible. There were other demands like the families of Hyde Park in the state of New York, that did a similar demand as Mrs. O'Hair in the year 1963, and another demand on the same year by Mr. Ed

Schempp in Philadelphia, PA. His was so that the reading of the bible is not allowed in schools. Organizations like these that place these types of demands even if they are local or state demand, when they reach the superior court, and they receive a ruling in their favor, they become national laws that affect all states. That was the case of the last two mentioned demands since Supreme Court. Joined both demands and voted 6-1 in favor of the plaintiffs, converting the law in a National Law. That is why we see lawyers, prosecutors, and judges use other demands and laws as guides in their decisions.

The liberals have achieved to remove prayers from schools since we can't even talk about it in the hallways. However, teaching about sex at an early age is completely allowed. The bible is not allowed but handing out birth control to young girls promoting sex, and avoiding pregnancy, is. There are schools that have already approved that if a minor is receiving birth control, they do not need to be informed or if she becomes pregnant and wants an abortion, the parents don't need to find out. The liberals will continue to fight against all Christian principles. It's the Democrats that have approved all of these ideals. It has come to a point that they now want to remove the "In God We Trust" from the money.

Everything that is happening started the following way. In 1802, President Thomas Jefferson, because of some meddling of the Federal Government against a Baptist Church in the state of Connecticut, through a letter they asked the Congress to legislate a law so that there could be a separation of the church and state. Clearly their intentions were to protect the practice of religion and their doctrines that are clear in the amendment that precisely the Congress did base on this petition. All of this changed when a liberal judge, Hugo Black, in the year 1947 interpreted that the intention of President Jefferson was to place a wall between the government and the church. This interpretation that until this day is under discussion, has been the base of the liberal Democrats to place the mentioned demands, and all other demands they have placed after this one. For example, they have placed many demands, which all have been granted, like remove crosses from certain places, paintings, forbid Christmas trees in governmental offices, including saying Merry Christmas, and even the mandates to the religious institutions of providing birth control pills to the women under the Obama Care law. This to say, it is unacceptable that a Christian put as primate political issues, personal benefits, and traditions, rather than spiritual matters. It is not in vain that the word of God says that many are called, but few are chosen. Every one that goes against the word of God should not receive our support and shouldn't even be our friends. The word of God in James 4:3-5,

> *3 And even when you ask, you don't get it because your motives are all wrong—you want only what will give you pleasure.*

> *4 You adulterers! Don't you realize that friendship with the world makes you an enemy of God? I say it again: If you want to be a friend of the world, you make yourself an enemy of God. 5 Do you think the Scriptures have no meaning? They say that God is passionate that the spirit he has placed within us should be faithful to him.[b]*

I can see with clarity how people continue to drift from all principles, and they are embracing everything that is against all what God has established. What hurst the most is watching Christians dance to the music they are being played. Mark 8:18 says they have eyes, but can't see, and they have ears but can't listen. I want to say that I have been in the Assembly Of God Churches in America, and the church nor the organizations, are in favor of any legislator that represents or support all the things here mentioned. Many Christians today seem to publicly show they do not fear God. I say they are clouds without water, dry rivers, and empty tanks.

> *Proverbs 25:14, ³ And even when you ask, you don't get it because your motives are all wrong—you want only what will give you pleasure.*
>
> *⁴ You adulterers! Don't you realize that friendship with the world makes you an enemy of God? I say it again: If you want to be a friend of the world, you make yourself an enemy of God. ⁵ Do you think the Scriptures have no meaning? They say that God is passionate that the spirit he has placed within us should be faithful to him.*

A true Christian seeks the eternal things and not the earthly ones.

> *Philippians 3:17-19, ¹⁷ Dear brothers and sisters, pattern your lives after mine, and learn from those who follow our example. ¹⁸ For I have told you often before, and I say it again with tears in my eyes, that there are many whose conduct shows they are really enemies of the cross of Christ. ¹⁹ They are headed for destruction. Their god is their appetite, they brag about shameful things, and they think only about this life here on earth.*

I have no doubts that in some cases the ignorance, and in other cases sin, have taken us to very dangerous times. I say this because there are many Christians that only think about the material things, and despite having the knowledge of who they are voting for, the material things seem to have more importance than the spiritual matters, and in even bigger cases, than the salvation of their souls. In other cases, there are Christians that don't even know who they are giving their votes to; they do it because they are in the political party they follow and of the influence of the liberal press. This will be the result of ignorance and the disobedience to God, as well as the things they are dragging from their childhoods.

I don't understand why many Christians don't seem to see that every time the Democrats are in majority in Congress, the liberals have been taking the country more and more against the things that God has established. If we observe the political platform of the Democrats and their beliefs together with their history, you don't have to be an expert to see what happens when they are in power. They are never good things when it comes to spiritual matters. There is no way people that are corrupt, mentally and spiritually, can give good fruit. For example, when President Bill Clinton ran for presidency, in his first term, there were approximately 800,000 homosexuals that the press covered, demanding their rights to be acknowledged. President Clinton told them that if he was elected, he would sign the law that acknowledged their rights, which he did immediately he was elected. After this law was signed, pastors have had to be very cautious in the way they preach against this sin. Anyone from this group can sew them. Don't think I am exaggerating. It has already happened.

On November 15, 2008, thanks to the law signed by this president, they presented in the Univision news of the state of California, thousands of lesbians and homosexuals demanding the state to allow them to get married. Embarrassing to see a father in this protest with a shirt that said: my daughter is lesbian, and I am her support. When they interviewed him, he said that his daughter had the right to be allowed to marry her partner. This is the result of voting and allowing liberals in Congress and the White House. We all know the story of ex-president Clinton who has been the most immoral president to have set foot in the White House. Not just for everything that was proven he did with Monica Lewinsky, but he also lied to the nation under oath, and is responsible for all the gay manifestations.

There are some Christians that are fooled by the press and politicians who with all the bad intentions, invent calumny, and in. most cases based on speculations. This is the reason why many Christians decide to vote for those candidates that have openly expressed themselves in favor of abortions and many other things that are against God's will. I still can't wrap my head around this fact and how many Christians can have peace with all of this. It is clear to me that the results will be catastrophic. It also proves the lack of knowledge and loyalty towards God. The ignorance is so big that there are Christians proud to say they are liberals. To be a liberal or to agree with them, is to say you are against God. How can a Christian agree with the Democratic Ideology? The word of God in 1 Timothy 5:22

> *22 Never be in a hurry about appointing a church leader.[a] Do not share in the sins of others. Keep yourself pure.*

Liberals call abortion and homosexualism part of the moral issues, but in reality, it is not as simple as the say it is. The bible calls it sin, and sin as worthy of death. This is not saying that someone is going to kill them; It is referring to eternal condemnation. They have earned hell with their acts.

> *2 Peter 3:5-7, 5 They deliberately forget that God made the heavens long ago by the word of his command, and he brought the earth out from the water and surrounded it with water. 6 Then he used the water to destroy the ancient world with a mighty flood 7 And by the same word, the present heavens and earth have been stored up for fire.*

They are being kept for the day of judgment when ungodly people will be destroyed.

Revelations 21:8, ⁸ *"But cowards, unbelievers, the corrupt, murderers, the immoral, those who practice witchcraft, idol worshipers, and all liars—their fate is in the fiery lake of burning sulfur. This is the second death."*

1 Corinthians 6: 9-15, ⁹ *Don't you realize that those who do wrong will not inherit the Kingdom of God? Don't fool yourselves. Those who indulge in sexual sin, or who worship idols, or commit adultery, or are male prostitutes, or practice homosexuality,* ¹⁰ *or are thieves, or greedy people, or drunkards, or are abusive, or cheat people— none of these will inherit the Kingdom of God.* ¹¹ *Some of you were once like that. But you were cleansed; you were made holy; you were made right with God by calling on the name of the Lord Jesus Christ and by the Spirit of our God.*

Avoiding Sexual Sin

¹² *You say, "I am allowed to do anything"—but not everything is good for you. And even though "I am allowed to do anything," I must not become a slave to anything. 13 You say, "Food was made for the stomach, and the stomach for food." (This is true, though someday God will do away with both of them.) But you can't say that our bodies were made for sexual immorality. They were made for the Lord, and the Lord cares about our bodies.* ¹⁴ *And God will raise us from the dead by his power, just as he raised our Lord from the dead.*

15 Don't you realize that your bodies are parts of Christ? Should a man take his body, which is part of Christ, and join it to a prostitute? Never!

Ephesians 5:6-17 says,

6 Don't be fooled by those who try to excuse these sins, for the anger of God will fall on all who disobey him. 7 Don't participate in the things these people do. 8 For once you were full of darkness, but now you have light from the Lord. So live as people of light! 9 For this light within you produces only what is good and right and true.

10 Carefully determine what pleases the Lord. 11 Take no part in the worthless deeds of evil and darkness; instead, expose them. 12 It is shameful even to talk about the things that ungodly people do in secret. 13 But their evil intentions will be exposed when the light shines on them, 14 for the light makes everything visible. This is why it is said,

"Awake, O sleeper,
rise up from the dead,
and Christ will give you light."

Living by the Spirit's Power

15 So be careful how you live. Don't live like fools, but like those who are wise. 16 Make the most of every opportunity in these evil days. 17 Don't act thoughtlessly but understand what the Lord wants you to do.

In 1 John 1:10,

10 If we claim we have not sinned, we are calling God a liar and showing that his word has no place in our hearts.

But one thing is to commit an involuntary sin, and another is to practice these sins, as stated in 1 John 3:8-10,

⁸ The wind blows wherever it wants. Just as you can hear the wind but can't tell where it comes from or where it is going, so you can't explain how people are born of the Spirit."

⁹ "How are these things possible?" Nicodemus asked.

¹⁰ Jesus replied, "You are a respected Jewish teacher, and yet you don't understand these things?

The sin of homosexualism the bible classifies them as shameful sins worthy of death. I have to remind you that because of this sin God reduced Sodom and Gomorrah and their surrounding cities, to ashes, which just as the others, having fornicated and gone after addictions, where placed in a place of eternal suffering. This is found in Jude 1:7

> *⁷ And don't forget Sodom and Gomorrah and their neighboring towns, which were filled with immorality and every kind of sexual perversion. Those cities were destroyed by fire and serve as a warning of the eternal fire of God's judgment.*

The word of God is very clear. That is why I don't understand the Christians that support these legislators that are against the word of God and are enemies of the Lord and are serving as instruments of Satan. Not so many years ago, this lifestyle wasn't acceptable in any society. But, in recent years, the devils has used corrupt people like liberal politicians such as Clinton and Obama, as well as famous people that are influencers, use their platforms, to take the message across in order for society to agree and accept the things that are against God. We are living the days where this and abortions is being promoted as something normal that society should accept. In a program of the Latin community that is very famous, I observed the following: the main purpose of the program was to make us understand that this is something normal that we have to understand. To the show they brought experts in the matter called psychologists, as well as a mom of a homosexual to educate society. To promote the acceptance of homosexualism, this mother stated that when she found out her son was gay, she was really affected, and that no mother agreed that their son could be gay. She said she was rebellious against this for eight years. Nevertheless, she said she had gone to therapy, and they helped her to see that there was nothing wrong with this. They alleged this was because of lack of education on her behalf and of all those that couldn't understand that this was normal. These types of expressions are very common in radio and television. This famous presenter interviewed the famous gay singer Ricky Martin, that was also interviewed by another reporter, where he said he was happy God made him like that. He also answered a question this presenter asked him: how can you explain that in the past you loved women, but now you love men? Ricky answered that God gave him the blessing of being able to love men and women. Counteracting this presenter, I sent him a letter explaining what the word of God says regarding this sin, how who practice it are an abomination to God, and how it is worthy of death.

The following information is very important to analyze. From our mother's womb we have inherited the germ of sin, not just homosexualism, but all sin. The excuse many people that defend them is that they are not guilty to have been born that way. Because of the following information, they have no excuse to been born that way or having turned that way. For example, there are homosexuals that say they were born that way and others say they were turned into that. Both cases are possible. We need to understand that in our mother's womb we inherit the genes of our ancestors. Some inherited traits from their parents and others from old relatives. That is why we sometimes say our children are identical to their fathers. Some inherit to be unfaithful, drunks, blasphemers, liars, deceivers, and even problematic. The list is long, but it is not in discussion if they were born or made like that. Regardless, that's why Christ came to die at the cross for all of our sins. The responsibility of all sinners is to repent from their behavior, accept the sacrifice Jesus made on the cross and change like the way Victor Manuelle's brother, a famous salsa singer from Puerto Rico, did. He practiced this behavior and was also interviewed by this famous presenter, and he gave a clear message of how God transformed him and proving what it has been established here. Like this testimony there are many more. Some of them I am aware of, and others have been presented by this same host in national television. The word of God in 1 Corinthians 6:9-11,

> *⁹ Don't you realize that those who do wrong will not inherit the Kingdom of God? Don't fool yourselves. Those who indulge in sexual sin, or who worship idols, or commit adultery, or are male prostitutes, or practice homosexuality, ¹⁰ or are thieves, or greedy people, or drunkards, or are abusive, or cheat people—none of these will inherit the Kingdom of God. ¹¹ Some of you were once like that. But you were cleansed; you were made holy; you were made right with God by calling on the name of the Lord Jesus Christ and by the Spirit of our God.*

It's a matter of repent and change.

Under the excuse that they were born this way means that abusers, thieves, unfaithful men, etc. shouldn't be blamed because, they were born this way.

The information of Ricky Martin saying God made him like that is completely wrong; even diabolic, or a big ignorance on his behalf. I question myself, where is the voice of the church? The church is allowing all of this attack and propaganda of darkness to spread, and there is no voice to step up and counterattack. I believe we need more brothers like Yiye Avila and Jorge Rasky. I know that for many, citing what the word of God says makes us discriminators or crazy fanatics. But I mention what is exactly written in his word. I don't discriminate against anyone. Like I tell my daughters, if anyone needs my help, I will help. The word of God tells us that God hates sin but loves sinners.

I know that the type of information I am sharing can result scandalous for many, because like I have said, the modern church is too comfortable. They are like the church of Revelations 3:15-22,

> 15 *"I know all the things you do, that you are neither hot nor cold. I wish that you were one or the other!* 16 *But since you are like lukewarm water, neither hot nor cold, I will spit you out of my mouth!* 17 *You say, 'I am rich. I have everything I want. I don't need a thing!' And you don't realize that you are wretched and miserable and poor and blind and naked.* 18 *So I advise you to buy gold from me—gold that has been purified by fire. Then you will be rich. Also buy white garments from me so you will not be shamed by your nakedness, and ointment for your eyes so you will be able to see.* 19 *I correct and discipline everyone I love. So be diligent and turn from your indifference.*

> 20 *"Look! I stand at the door and knock. If you hear my voice and open the door, I will come in, and we will share a meal together as friends. 21 Those who are victorious will*

sit with me on my throne, just as I was victorious and sat with my Father on his throne.

[22] *"Anyone with ears to hear must listen to the Spirit and understand what he is saying to the churches."*

Abortions and homosexualism are a practice of sin, so anyone that supports them is going against the word of God. You can use any type of excuse to give these legislators a vote, but there is only one reality. If liberal candidates were not elected, they wouldn't have legislated laws that allowed for millions of innocents to be killed and until God's punishment doesn't come down upon them, they will continue to kill. The Christians that are aware of these things will have to come before God and give their explanations of why they supported these candidates, going after personal benefits or following traditions. We have to analyze correctly what Philippians 3:18-19 says:

[18] *For I have told you often before, and I say it again with tears in my eyes, that there are many whose conduct shows they are really enemies of the cross of Christ.* [19] *They are headed for destruction. Their god is their appetite, they brag about shameful things, and they think only about this life here on earth.*

Let's analyze the following: if a person knows another person's intentions and they facilitate them a weapon, and they kill somebody, are they not participants of that crime? There isn't a difference. If we are aware that we are going to vote in favor of someone that has already said they are in favor of abortions, homosexualism, and many other sins, aren't we directly or indirectly participants of that sin? Can we use the excuse that many say: that's their problem! As per the word of God, there is no excuse. For example, President Obama always said he was in favor of abortions, same sex marriage, cloning, and many other sins like most of the Democrats. Like I mentioned, because of the lack of knowledge and disobedience, we must live difficult and catastrophic times that we supported ourselves. For example, at the beginning of 2000, because the Democrats had approved same sex marriage in 5 states, and they were trying to convert it into a national law, so they could get married in all states, by petition of all religious organizations, they asked the Congress that they amend the Constitution or to create a law that marriage should only be between a man and a woman. They tried to do the amendment, but the Democrats opposed. The Republicans tried through a law, and the law was approved in 2004, signed by GW Bush, known as DOMA. During the years that Bush was the president, the subject of these marriages ended. But, when President Obama took the presidency, he declared he wasn't in favor of this law and that he didn't support it. That is how he opened the door to the homosexuals who took advantage of his postures, taking the case again to the supreme court, who voted 5-4 in favor of these marriages. I am more than sure that all of the Christians who voted for Obama, have not noticed the mistake they made electing him president. In the days we are living, the Democrats are at war against President Trump for his nominations of conservative Judges. That was the reason they went against Judge Brett Kavanaugh with false accusations and doing everything possible so that he wouldn't reach the Supreme Court. They alleged this could be a threat in all of the advancements they had conquered on the things previously mentioned. Now they are trying to do the same thing with conservative Judge Amy Coney Barrett, because the Democrats allege that her religious beliefs, she is a threat for the liberals. For example, in an interview that the California Senator Diane Feinstein, questioned her doctrines in abortions, same sex marriage, telling her that her doctrines are very worrisome for a Judge of the Supreme

Court. This leaves clear the intentions and the purpose of the Democrats.

There is no doubt, and the records show that the Democrats are responsible of approving the following laws and beliefs:

1. Abortions

2. Same Sex Marriage

3. Prohibitions of talking about God or praying in the schools

4. Eliminate the Bible from the courthouses and schools

5. Sex education in school since very early age

6. Birth control for underage minors without their parents consent in order to avoid pregnancy.

7. Homosexual demonstrations anywhere they want, but Christians can't talk about God at school

8. Nominations of liberal judges under the influences of the Democrats, not only at the Supreme Court but at state level too. They have also made decisions in favor of demands that have been submitted to not allow Christmas Trees, and Christmas decorations in government offices like libraries and other public places. And many other demand that have been mentioned and for sure they will continue to legislate against everything God has established. That is why they are so opposed to conservative judges.

At the beginning I informed that depending on the posture, traditions and culture, this information would be difficult to swallow, and in many cases accepted. I believe it is time to start raising our voice and say things as it is, even if it hurts. When parents discipline their kids, it's not because they don't love them. On the contrary, they do it because they love them and want the best for them. This is the purpose for all these truths. I love both Christians and sinners, trying to correct what is wrong is just a demonstration of love.

> *Hebrews 12:4-6,* **⁴ *After all, you have not yet given your lives in your struggle against sin.***
>
> **⁵ *And have you forgotten the encouraging words God spoke to you as his children?[a] He said,***
>
> **"*My child,[b] don't make light of the LORD's discipline, and don't give up when he corrects you.***
> **⁶ *For the LORD disciplines those he loves, and he punishes each one he accepts as his child.***

> *Colossians:2-8,* **⁸ *Don't let anyone capture you with empty philosophies and high-sounding nonsense that come from human thinking and from the spiritual powers[a] of this world, rather than from Christ.***

What I advise to citizens, especially to Christians is not to certify as facts what the press puts out. The press, social media, internet webpages, are deceiving people and injecting venom 24-7. They are the instruments the devil is using to deceive humanity. For example, a Pastor wrote in Facebook that he didn't know what to think about the conservatives of this country, because to Martin L. King the conservatives placed recorders to their phones accusing him of communist and they persecuted him. Know we know that the people behind this was the Kennedy administration, who had given his brother Robert, who was the AG of the nation, the order. The order were done in some meetings that MLK had with communists. And I should clear up that to MLK nobody proved anything wrong against him. But it was the Kennedy's that accused him of communist and not the conservatives of the country. When I asked the pastor where he obtained that information, he said he got it from Google. On a daily basis, people and Christians take as facts what the press and social media put out. It seems that they are easily deceived. The media today are not true journalists, they are activists of the political parties. Christians have to be careful not to go off based on what other people think or say, without having the correct information. The word of God says the following in Romans 2:1-6,

> *You may think you can condemn such people, but you are just as bad, and you have no excuse! When you say they are wicked and should be punished, you are condemning yourself, for you who judge others do these very same things. [2] And we know that God, in his justice, will punish anyone who does such things. [3] Since you judge others for doing these things, why do you think you can avoid God's judgment when you do the same things? [4] Don't you see how wonderfully kind, tolerant, and patient God is with you? Does this mean nothing to you? Can't you see that his kindness is intended to turn you from your sin?*

[5] But because you are stubborn and refuse to turn from your sin, you are storing up terrible punishment for yourself. For a day of anger is coming when God's righteous judgment will be revealed. 6 He will judge everyone according to what they have done.

Why do I share this information? Through social media there are many Christians that constantly are speaking atrocities and accusations against President Trump. They accuse of racism, liar, deceit, criminal, and many more things including trying to fool Christians. They justify their accusations with bible verses taken out of context. For example, someone wrote to me in a post I wrote in Facebook, defending the things I just wrote about, that the word of

God says that we must love our neighbors. (He forgot what the word says love our enemies and do good to those that despise us). It continues to say that we must care for the orphans, widows, poor, and immigrants. As followers we can't stay with our hands crossed since this president continues to kill many innocent people. This is the information is from CNN and other liberals. Biblically it is impossible that from a same fountain, obtain sweet water and sour water.

> *James 3:10,* [10] *And so blessing and cursing come pouring out of the same mouth. Surely, my brothers and sisters, this is not right!*

> *Ephesians 4:29- 32,* [29] *Don't use foul or abusive language. Let everything you say be good and helpful, so that your words will be an encouragement to those who hear them.*

> [30] *And do not bring sorrow to God's Holy Spirit by the way you live. Remember, he has identified you as his own,[a] guaranteeing that you will be saved on the day of redemption.*

> [31] *Get rid of all bitterness, rage, anger, harsh words, and slander, as well as all types of evil behavior.* [32] *Instead, be kind to each other, tenderhearted, forgiving one another, just as God through Christ has forgiven you.*

> *Romans 12:14-21,* [14] *Bless those who persecute you. Don't curse them; pray that God will bless them.* [15] *Be happy with those who are happy, and weep with those who weep.* [16] *Live in harmony with each other. Don't be too proud to enjoy the company of ordinary people. And don't think you know it all!*

[17] Never pay back evil with more evil. Do things in such a way that everyone can see you are honorable. [18] Do all that you can to live in peace with everyone.

[19] Dear friends, never take revenge. Leave that to the righteous anger of God. For the Scriptures say,

"I will take revenge;
  I will pay them back,"[a]
  says the Lord.

[20] Instead,

"If your enemies are hungry, feed them.
  If they are thirsty, give them something to drink.
In doing this, you will heap
  burning coals of shame on their heads."[b]

[21] Don't let evil conquer you but conquer evil by doing good.

The way many express themselves, we don't have to be experts to see that their hearts are filled with hate and rebellion and are being dominated by these feelings.

James 3:13-18, [13] If you are wise and understand God's ways, prove it by living an honorable life, doing good works with the humility that comes from wisdom. [14] But if you are bitterly jealous and there is selfish ambition in your heart, don't cover up the truth with boasting and lying. [15] For jealousy and selfishness are not God's kind of wisdom. Such things are earthly, unspiritual, and demonic. [16] For wherever there is jealousy and selfish ambition, there you will find disorder and evil of every kind.

[17] But the wisdom from above is first of all pure. It is also peace loving, gentle at all times, and willing to yield to others. It is full

of mercy and the fruit of good deeds. It shows no favoritism and is always sincere. [18] And those who are peacemakers will plant seeds of peace and reap a harvest of righteousness.

Some people believe that Trump is deceiving Christians, but they haven't noticed that the ones being deceived by the liberal press are them. The worst part is that all this behavior is to defend the politicians that are enemies of God and the church. They are attacking the person that is in favor of what God has established and has shown it with his words and actions. I will ask you again, how can a Christian approve of a politician that is in favor of abortions, same sex marriage, the prohibition of bible and prayers in schools, amongst many other things? And then use the word of God to justify their actions. I say it because the way they quote the bible, mostly out of context, the seem not to know what the word in Colossians 1:9 teaches us. Having knowledge and spiritual wisdom is one thing, and having biblical wisdom is another. Biblical knowledge can be intellectual; the biblical even though it is spiritual, can be more profound. He who has the spiritual knowledge is guided in everything he does and says by the spirit. This is only achieved through repentance and the continuous search of God.

> *Romans 8:1-8, So now there is no condemnation for those who belong to Christ Jesus. 2 And because you belong to him, the power[a] of the life-giving Spirit has freed you[b] from the power of sin that leads to death. 3 The law of Moses was unable to save us because of the weakness of our sinful nature.[c] So God did what the law could not do. He sent his own Son in a body like the bodies we sinners have. And in that body God declared an end to sin's control over us by giving his Son as a sacrifice for our sins. 4 He did this so that the just requirement of the law would be*

fully satisfied for us, who no longer follow our sinful nature but instead follow the Spirit.

5 Those who are dominated by the sinful nature think about sinful things, but those who are controlled by the Holy Spirit think about things that please the Spirit. 6 So letting your sinful nature control your mind leads to death. But letting the Spirit control your mind leads to life and peace. 7 For the sinful nature is always hostile to God. It never did obey God's laws, and it never will. 8 That's why those who are still under the control of their sinful nature can never please God.

Galatians 5:16, 16 So I say, let the Holy Spirit guide your lives. Then you won't be doing what your sinful nature craves.

THE IMPORTANCE OF KNOWING IN WHAT CHURCH TO CONGREGATE

Like I have said before, this behavior only confirms we are living in the end of times. Christ warned us not to be deceived by no one, even those that preach the gospel and call themselves men and women of God. The following may sound hard for some people, but if it weren't a reality that is hurting the spiritual life of many people, I wouldn't mention it.

It's been a while that we are seeing how a worrisome group of pastors, evangelists, and prophets, are taking the gospel of the Lord and the pulpits, to teach and practice doctrines of error or false. Some due to lack of knowledge, and others to obtain personal benefits and get enriched of a great number of Christians that don't have the biblical knowledge and are being deceived. There are people driven by their eloquence, their emotions, abilities, and charisma. They blindly believe everything the speak. Clearly the warning that Christ left in Matthew 24:4-5 is being fulfilled.

> *4 Jesus answered: "Watch out that no one deceives you. 5 For many will come in my name, claiming, 'I am the Messiah,' and will deceive many.*

> *Hosea 4:6, my people are destroyed from lack of knowledge. "Because you have rejected knowledge, I also reject you as my priests; because you have ignored the law of your God, I also will ignore your children.*

The knowledge I am referring to is the spiritual knowledge:

> *Colossians 1:9 For this reason, since the day we heard about you, we have not stopped praying for you. We continually ask God to fill you with the knowledge of his will through all the wisdom and understanding that the Spirit gives,*

Since a few years back, certain personal experiences taught me the importance of having knowledge of absolutely everything before making decisions. That is the reason that no matter what the cause is, I analyze everything carefully, verify all fountains, and make sure to have the correct information. The lack of knowledge is what destroys many people and even entire cities. Everything informed here is backed up by the word of God with its bible references, as well as clear information from theologians and ministers of God. Since I was born I have been in church, and even though there are many Christians that have been in church for years, and are still drinking milk and are not eating solid foods of the word of God. In my case I had a mother that was very spiritual, teacher at the Biblical Institute; like my dad who was another great teacher, she was a missionary for more than 50 years. They took charge of educating and doctrine us in the word of God. Out of those teachings, 3 of the 5 sons are pastors. Joel and Jose both have doctorates in Theology and in psychology. My sister Sara has been a minister for more than 40 years. This information is not for our glory, because Apostle Paul said in 2 Corinthians 10:17-18,

17 But, "Let the one who boasts boast in the Lord."[a] 18 For it is not the one who commends himself who is approved, but the one whom the Lord commends.

I give this information because it is important to know the fountain that gives the information. Like I said, it's the disturbing behavior that ministers, as well as the congregation, that have taken me to study profoundly the following information. My purpose is the same as Paul said in 2 Corinthians 10:3-5:

3 For though we live in the world, we do not wage war as the world does. 4 The weapons we fight with are not the weapons of the world. On the contrary, they have divine power to demolish strongholds. 5 We demolish arguments and every pretension that sets itself up against the knowledge of God, and we take captive every thought to make it obedient to Christ.

This is when I will speak on the importance of knowing in what church we congregate at. There is a hymn that is sung at Evangelical Pentecostal Churches that says: It doesn't matter the church that I go to, if you are the one behind the Calvary. If your heart is like mine, give me your hand and my brother you will be. Even though the writer said: if you are behind the Calvary, I think we should care about the church we assist or visit. Further along I inform the strong reasons; so powerful that it can depend on us being saved.

Our Savior Jesus Christ is the founder of our church. Make note that I speak on one church that is His. As we know, He left his Throne of Glory, and became human going through a process, until He was crucified, to give us salvation, those that accept Him as their savior, and live a life according to his commandments. Part of this process was teaching his disciples, and prepare them, so after his ascendance to heaven back to this throne of glory, they would continue his ministry and win more souls for his kingdom. He went back to heaven and promised them the Holy Spirit to be with them and his church. Acts 1:8 But you will receive power when the Holy Spirit comes on you; and you will be my witnesses in Jerusalem, and in all Judea and Samaria, and to the ends of the earth." We find that on the day of Pentecostal, He arrived as it was promised by Jesus, chapter 2 of the same book, that was the day that the church started to exist and was anointed.

How do we describe the Church of Christ?

Myer Pearlman describes it the following way: (I quote) What is the church? We could respond the question considering the following: (1) the voices that describe this institution; (2) the voices that describe Christians; (3) the illustrations that describe the church.

The voices that describe the church: the Greek voice, New Testament, to describe the church is "Ekklesia" that in Spanish means "assembly of the called". The term is applied to the whole body of Christians, to a congregation, to the body of believers of all of the earth. The church is a brotherhood or spiritual communion, in which all of the differences that separate humanity has been abolished. "There are no Jew or Greek"; It surpasses the most profound of the divisions, based on the religious history; there is no slave or free; it surpasses the most profound of the cultural divisions; there is no male or female; and it surpasses the most profound of human divisions.

The church is an organism and not merely an organization. An organization is a group of people congregated voluntarily for a certain purpose, just like a fraternal organization or a syndicate. An organism is something alive that develops from the life within. In a figurative sense, it is the total sum of all the related parts, in which the relation of each of the parts encloses a relationship with it All. A car can be denominated an organization of certain parts or mechanical pieces. A human body is an organism since it is composed of many members and animated common organs. The church of Christ is a body composed of millions of born again Christians.

Christ didn't come to establish religions, but one religion: His religion. Which is this religion? Accepting Him as the only Savior and live a life separated for Him and following His commandments. Which commandments? He left us his Holy Word in 66 books, written by 40 authors, which God inspired to write. In these books we find laws, mandates, and his doctrines. After the day of Pentecostal, the Apostle Peter preached his first sermon. The Apostles continued preaching, teaching, and baptizing in the name of Jesus Christ as He ordered them. It was necessary to build churches that only preach what He taught and ordered us.

> *James 1:26-27* [26] *Those who consider themselves religious and yet do not keep a tight rein on their tongues deceive themselves, and their religion is worthless.* [27] *Religion that God our Father accepts as pure and faultless is this: to look after orphans and widows in their distress and to keep oneself from being polluted by the world.*

The religions have been formed on a later date. The Catholic church claims to be the first since it was founded 30 years after Christ. The Orthodox 988 in Russia. The Church of Jesus Chris in 1830. The evangelicals are a variety of churches leaded by Martin Luther in 1517. Martin Luther was a university teacher and catholic, didn't agree and he publicly opposed to the sale of indulgences of the church. The indulgences was a document issued by the Catholic Church, that avoided having to complete a penance or punishment for the sins committed. Also, to reduce the stay at the purgatory once they died; reason why the Catholic church ex-communed him.

The Evangelicals are a mix of churches with different origins, beliefs, and ways of organizing. They are churches of denominations.

The Lutherans are followers of Martin Luther that was established in 1517.

The Presbyterians were founded in the 16th century.

Baptists in the 17th.

Methodists in the 18th

Pentecostals in the 19th

Jehovah Witness were founded in 1870 in Pittsburg, PA.

Mormons in 1838 in Salt Lake City. Neither of these religions are evangelicals.

The Adventists descendants of the Pentecostals, have a belief and doctrine completely different to the Pentecostals and other evangelicals.

Pentecostals which was one of the most recent denominations, replicated quickly in the 20th century, being one of the fastest growing denominations. In the Pentecostals there are multiple subdivisions. I am not going to mention them all, but some examples are: The Disciples of Christ, Church of God, Christian Assembly, Defenders of Faith, Christian Alliance and Missionary, Prophets, and many others, including a large number of independent churches.

The Pentecostals are classified as classic conservatives, and other neo-Pentecostals are more modern. We can use as an example, the Pentecostals of MI (International Movement in Spanish), as the more classic and the Assembly of God as the modern, even though they are also conservative.

The reason these religions took so long to be established is because the Bible was written in Hebrew and in Aramaic, and it was prohibited to make any translations of it. For example, it was translated to Spanish in the year 1530.

These are some of the reasons why I had mentioned it was important to know in what church we decide to congregate. I start with the following personal experience.

At the beginning of the 80's, a young Catholic man that was my sister-in-law's boyfriend told me he needed to talk to me and clarify many things he didn't understand of us Pentecostals. We met at a restaurant, and he told me the following: I am a good Catholic, and I go every week to church. I am not a Catholic by name, and I just don't understand certain things of the Pentecostals, and they are causing me problems with my girlfriend. What I am going to ask you has to have a convincing answer. If you convince me with your response, I assure you that next Sunday, I will convert to Pentecostal. He continued to say, what I am going to ask you is simple, but I don't know what explanation you can give me. What sin is there in going to the beach with your girlfriend? Because her family doesn't allow it. He finished saying; I don't think that anywhere in the Bible it says this is a sin. In all honesty, it wasn't what I expected. Until this day I believed God illuminated me to give him this response. As you may know, the purpose of all religions, excluding some cults, like the Rosicrucian, the spiritisms, sinters and others; but in general, the churches have as a primary purpose, and hat is to take us closer to God so we can live a life according to His word, and we can be saved. They achieve this through their doctrines and teachings. He agreed. So, the responsibility of a person that wants to live a life closer to God, and try to sin the least possible, they examine which religion has the best teachings and doctrines. The answer to your question is the following: Bathing in the beach is not a sin. The reason Pentecostals prohibit bathing at the beach is because it doesn't matter how Christian you are, when you go to the beach and there are people with very little clothing, and in many cases exposing it all, it takes you to the sin of greed and it corrupts your mind. Like I explained, the Pentecostal church wants to avoid its members to sin against God. His response was he had no arguments. The following week, he kneeled on the altar.

Like I mentioned, every church has its beliefs and doctrines. The same Evangelical Pentecostal churches are sovereign, and affiliated to the same council, even though they have the same doctrine, they vary in certain beliefs, depending on their leaders, they can even establish some dogmas and some of them don't adjust to the word of God. I am going to give an example.

There are pastors and evangelicals that believe and teach that a Christian can't suffer depression. Belief that I do not share, and I also base on the word of God. Through the bible we find certain characters that in moments of adversity, get depressed. In some cases, they even asked God to take their lives.

The difference is that a true Christian, that knows the word of God, knows how to reject all darts launched by Satan, through prayer and faith in God, which sustains us. There are also some Christians that Satan catches them off guard since they have their house founded on sand, and other years serving the Lord without any roots. These types of Christians are an easy target for Satan. They take steps back from serving the Lord and accepting Satan's temptations.

In First of Samuel, we find a woman clearly depressed. We find a man named Elkanah had two women. His primary wife names Hannah and the other Peninnah. Peninnah was able to give Elkanah children, but Hannah wasn't able to, although it was her desire. The story says that every year Elkanah offered sacrifices to Jehovah, and he would give Peninnah and his children their part. To Hannah we would give a chosen part because she didn't have any children. Peninnah would bully her and upset het because God hadn't let her conceive. The story narrates that Hannah was so sad that she didn't eat, and she was so heartbroken that she cried very sourly. When the priest Eli saw the condition, she was in, he saw her so depressed he thought Hannah was drunk. But Hannah answered Eli saying: No, my Lord, I am not drunk, I am a troubled woman in the spirit. Because at the magnitude of my sadness and my affliction I have spoken now. When the Priest Eli told Hannah God would let her have a son, immediately what we know today as depression, disappeared. She ate and drank and was no longer afflicted.

In 1 Kings 19, we find the Prophet Elijah after he had defeated the 400 prophets of Baal, after he showed them who the true God was, and after he slaughtered them, now we find him running away, since King Ahab gave Jezebel the news of what Elijah had done, Jezebel sent him a message to consider himself a dead man by the next day the latest. Elijah ran away, and seeing the condition he was in, he got depressed to the point of asking God to take his life.

In the book of Job, on chapter 3, we also find a depressed Job because of his sickness. This is what Job said:

After this, Job opened his mouth and cursed the day of his birth. ² He said:

³ "May the day of my birth perish, and the night that said, 'A boy is conceived!' ⁴ That day—may it turn to darkness; may God above not care about it; may no light shine on it. May gloom and utter darkness claim it once more; may a cloud settle over it; may blackness overwhelm it.

⁶ That night—may thick darkness seize it; may it not be included among the days of the year nor be entered in any of the months. May that night be barren; may no shout of joy be heard in it. ⁸ May those who curse days[a] curse that day, those who are ready to rouse Leviathan.

⁹ May its morning stars become dark; may it wait for daylight in vain and not see the first rays of dawn, for it did not shut the doors of the womb on me to hide trouble from my eyes. ¹¹ "Why did I not perish at birth, and die as I came from the womb? ¹² Why were there knees to receive me and breasts that I might be nursed? ¹³ For now I would be lying down in peace; I would be asleep and at rest ᵂⁱᵗʰ kings and rulers of the earth, who built for themselves places now lying-in ruins, ¹⁵ with princes who had gold, who filled their houses with silver. ¹⁶ Or why was I not hidden away in the ground like a stillborn child, like an infant who never saw the light of day?

¹⁷ There the wicked cease from turmoil, and there the weary are at rest. Captives also enjoy their ease; they no longer hear the slave driver's shout. ¹⁹ The small and the great are there, and the slaves are freed from their owners. ²⁰ "Why is light given to those in misery, and life to the bitter of soul, ²¹ to those who long for death that does not come, who search for it more than for hidden treasure, ²² who are filled with gladness and rejoice when they reach the grave? ²³ Why is life given to a man whose way is hidden, whom God has hedged in?

²⁴ For sighing has become my daily food; my groans pour out like water. ²⁵ What I feared has come upon me; what I dreaded has happened to me. ^{26} *I have no peace, no quietness; I have no rest, but only turmoil."*

If Job would have expressed himself like that in front of Christians that think that people who talk like that are possessed by demons of darkness, they would have tried to rebuke him immediately. Job didn't need deliverance from darkness, as soon as God healed him and he was ok, the sadness disappeared. I ask myself; how many people have had this illness and they have been prayed on, and prayed on, rebuking demons and spirits from darkness? My honest answer is that because of the lack of knowledge I have seen these prayers since I was young.

There are also Christians that firmly believe and preach that a Christian can't get sick. They base it on Isaiah 53:4-5,

> *Surely, he took up our pain and bore our suffering, yet we considered him punished by God, stricken by him, and afflicted.* ^{5} *But he was pierced for our transgressions, he was crushed for our iniquities; the punishment that brought us peace was on him, and by his wounds we are healed.*

On one occasion, a brother I knew really well told me: I know a good brother that is a Pastor. As you may know, there aren't any Hispanic Works in this area, and we agreed that I would lend him my house so that he can start a church. Since we know each other, I would love it if you and your daughters can help us with the music and worship. I agreed. Like a month later the Pastor preached a message and emphatically said in the message that a Christian that confessed to be sick was offending God because in Isaiah 53:4-5 clearly says that He already carried our sickness and we have been cured. When we went out, I took him aside along with the brother that had asked us to help. I told him that based on the word of God he was giving a wrong information in reference to a Christian getting sick. The brother got really upset and told me: Brother Del Toro, you just blasphemed against God. What you just said is making God a liar. The word we just read says we have already been cured. How do you dare counteract against the word of God? I told him I wasn't counteracting against the word of God, and I was trying to explain. But the brother was so upset that he told me that he no longer wanted to listen to me. The other brother asked him to allow me to expose my point of view. He told him he really knew me and that I had a lot of knowledge of the word. The Pastor didn't want to hear me because he said there was nothing, I could say that could contradict what the word of God said. After a while he agreed to listen. I asked him: Do you consider knowing about the word of God than the Apostle Paul? He answered no. So, let's see what the Apostle has to say regarding this matter.

> *Philippians 2:19 -30; 9 I hope in the Lord Jesus to send Timothy to you soon, that I also may be cheered when I receive news about you. 20 I have no one else like him, who will show genuine concern for your welfare. 21 For everyone looks out for their own interests, not those of Jesus Christ. 22 But you know that Timothy has proved himself, because as a son with his father he has served with me in the work of the gospel. 23 I hope, therefore, to send him as soon as I see how things go with me. 24 And I am confident in the Lord that I myself will come soon.*
>
> *25 But I think it is necessary to send back to you Epaphroditus, my brother, co-worker, and fellow soldier, who is also your messenger, whom you sent to take care of my needs. 26 For he longs for all of you and is distressed because you heard he was ill. 27 Indeed he was ill, and*

almost died. But God had mercy on him, and not on him only but also on me, to spare me sorrow upon sorrow. 28 Therefore I am all the more eager to send him, so that when you see him again you may be glad and I may have less anxiety. 29 So then, welcome him in the Lord with great joy, and honor people like him, 30 because he almost died for the work of Christ. He risked his life to make up for the help you yourselves could not give me.

1 Timothy 5:22, Do not be hasty in the laying on of hands, and do not share in the sins of others. Keep yourself pure.

Galatians 4:13, As you know, it was because of an illness that I first preached the gospel to you,

After I gave him all of these undeniable verses he said, Why then, does Isaiah say we have already been cured? The response is clear in Matthew 8:14-17,

14 When Jesus came into Peter's house, he saw Peter's mother-in-law lying in bed with a fever. 15 He touched her hand and the fever left her, and she got up and began to wait on him.

16 When evening came, many who were demon-possessed were brought to him, and he drove out the spirits with a word and healed all the sick. 17 This was to fulfill what was spoken through the prophet Isaiah:

*"He took up our infirmities
and bore our diseases."*

The explanation to all of this is very simple. The word also told us that Christ dies to save the sinners: Does this mean that everyone is already saved? Absolutely not. It is a provision Christ offers through the cross for everyone that believes in Him, accepts Him, and serves

Him. The same way Isaiah 53 is a provision for all of those that have faith and asks God for healing, through the sacrifice on the cross, they can receive it.

There are also pastor and evangelists that preach about the super faith. They believe a person can get sick but has to declare they are healed, even if they still feel pain. If they can't barely walk, and he have herniated discs, they can't straighten; they have to walk as if all is good. The word teaches us to have faith the size of a mustard seed, to pray to God with faith, and healings are all based on having faith. But saying we are healed while still being sick is simply becoming liars. We have to stimulate to have faith, but not under lies. This matter about faith and healings has many people confused. There are people that blame God for their sickness or the deaths of family members. There are cases like the case of Job, my mother's, and other cases that God doesn't send them the sickness but allows them. John 11:4 says: *But when Jesus heard about it, he said, "Lazarus's sickness will not end in death. No, it happened for the glory of God so that the Son of God will receive glory from this."* I think it's very important to understand that all humans were created to be eternal. But our lives are divided into two parts. When God created Adam and Eve, He created them to live eternally on earth, but sin took them to death. People in the past lived for more than 500 years. Methuselah was the one that lasted the most, 969 years. Because of sin and man's bad behavior, God reduced the years to 120 years on earth. Man was and has been disobedience and, in His word, God said: Psalm 90:7-10,

> *We are consumed by your anger and terrified by your indignation. ⁸ You have set our iniquities before you, our secret sins in the light of your presence. ⁹ All our days pass away under your wrath; we finish our years with a moan.*

¹⁰ Our days may come to seventy years, or eighty, if our strength endures; yet the best of them is but trouble and sorrow, for they quickly pass, and we fly away.

Because of our earthly nature, our bodies get sick. From birth we drag the sickness germs, we also get sick because of what we eat, when we eat the wrong things and in occasions, we consume venom things or with bacteria, or too much grease, among other things. My dad was a Christian man for many years. He lived an entire life without addictions. He got sick with cancer and died. In his funeral a neighbor made the following comment: I don't understand how a man like your dad without any addictions and Christian can die that way. What he didn't know is that my dad worked for many years with asbestos, and the doctor's report said that caused him the cancer. I read in a science book that there are many herbs that people use to eat or nourish, that little by little poison their neurons and as a result, causes different mental illness. On the other hand, it doesn't matter how much we take care of ourselves, it has already been established that we are going to die. In the case of early age at 80 and a longer age at 100. The questions that some ask is: why does the bible say the more robust 80 and some last 100? The bible also says that if you honor your parents, your days on earth will be extended. But we will surely leave. That is the first part already mentioned of the eternity. From there we pass to the glory or to hell to wait for the grand jury of the White Throne of Revelations 20, for eternal condemnation, or enjoy with God for eternity. Whoever is submitted to God and feels sure of his outcomes, doesn't fear death. That is why the Apostle Paul said in Philippians 1:21

For to me, to live is Christ and to die is gain.

Jesus himself in John 11:25-26 said:

25 Jesus said to her, "I am the resurrection and the life. The one who believes in me will live, even though they die; 26 and whoever lives by believing in me will never die. Do you believe this?"

Another example:

This information is regarding a doctrine practice of the *"Josué Assemblies of God"* church in Broward, Florida whose mother church is located in San Salvador, El Salvador. I want to make it clear that the doctrine I am about to speak on is exclusively to this particular church and not its council. This is just another example of what I have mentioned in the differences between churches.

The doctrine is the following: they believe that a person after having converted, and even if they are prepared to leave with the Lord, they could have darkness spirits therefore, they need to be delivered. They deliver them in the following manner: the pastor or the spiritually prepared for this exorcism or deliverance are in prayer and fastening for a week. Then they take the person aside and ask them about everything negative that has happened to them since they were born. Questions like if they have any grudges with anyone, if they have been abused by anyone, and because of that they carry grudges, if they are jealous, if they hate someone, if they have been cursed by someone or practice spiritism. All types of things that can be holding someone back from enjoying a happier and better life, both spiritually and mentally. After they do the prayer rebuking all the spirits in line with the information they have obtained. The prayer lasts for hours.

This practice is not in line with the word of God. The following references teach the contrary.

2 Corinthians 5:17 Therefore, if anyone is in Christ, the new creation has come:[a] The old has gone, the new is here!

1 Corinthians 3:16, Don't you know that you yourselves are God's temple and that God's Spirit dwells in your midst?

John 14:23, 23 Jesus replied, "Anyone who loves me will obey my teaching. My Father will love them, and we will come to them and make our home with them.

Romans 8:9-11, 9 You, however, are not in the realm of the flesh but are in the realm of the Spirit, if indeed the Spirit of God lives in you. And if anyone does not have the Spirit of Christ, they do not belong to Christ. 10 But if Christ is in you, then even though your body is subject to death because of sin, the Spirit gives life[a] because of righteousness. 11 And if the Spirit of him who raised Jesus from the dead is living in you, he who raised Christ from the dead will also give life to your mortal bodies because of[b] his Spirit who lives in you.

2 Corinthians 3:17, Now the Lord is the Spirit, and where the Spirit of the Lord is, there is freedom.

It is clear that where the Spirit of God lives, no other spirit can cohabitate. The enemy can't touch anyone that has been born from the Lord.

1 John 5:18, We know that anyone born of God does not continue to sin; the One who was born of God keeps them safe, and the evil one cannot harm them.

1 Corinthians 6:17-20, But whoever is united with the Lord is one with him in spirit. 18 Flee from sexual immorality. All other sins a person commits are outside the body, but whoever sins sexually, sins against their own body. 19 Do you not know that your bodies are temples of the Holy Spirit, who is in you, whom you have received from God? You are not your own; 20 you were bought at a price. Therefore, honor God with your bodies.

When a person repents and is born from God, immediately they are justified, regenerated, and sanctified.

1 Corinthians 6:11, And that is what some of you were. But you were washed, you were sanctified, you were justified in the name of the Lord Jesus Christ and by the Spirit of our God.

Titus 3:5, he saved us, not because of righteous things we had done, but because of his mercy. He saved us through the washing of rebirth and renewal by the Holy Spirit,

Of course, once we are saved, we begin a process of renovation and cleanliness, that will not end until we are with the Lord.

Colossians 3:5-10, Put to death, therefore, whatever belongs to your earthly nature: sexual immorality, impurity, lust, evil desires, and greed, which is idolatry. 6 Because of these, the wrath of God is coming.[a] 7 You used to walk in these ways, in the life you once lived. 8 But now you must also rid yourselves of all such things as these: anger, rage, malice, slander, and filthy language from your lips. 9 Do not lie to each other, since you have taken off your old self with its practices 10 and have put on the new self, which is being renewed in knowledge in the image of its Creator.

I want to clarify that if after being saved a person slips and starts to practice sin or start to visit places that opens the door to Satan, one of those spirits can come over them since the Holy Ghost can't coexist with sin. That is why the Lord said in Matthew 12:43-45,

"When an impure spirit comes out of a person, it goes through arid places seeking rest and does not find it. 44 Then it says, 'I will return to the house I left.' When it arrives, it finds the house unoccupied, swept clean and put in order. 45 Then it goes and takes with it seven other spirits more wicked than itself, and they go in and live there. And the final condition of that person is worse than the first. That is how it will be with this wicked generation."

These filthy spirits that it is referring to are demons. I want to clear up that to commit a sin is one thing but to practice sin is another. This is clear in 1 John 1:8,

If we claim to be without sin, we deceive ourselves and the truth is not in us.

1 John 3:8, The one who does what is sinful is of the devil, because the devil has been sinning from the beginning. The reason the Son of God appeared was to destroy the devil's work.

For the reason already mentioned, a person can be possessed by an evil spirit, but not in the context this church teaches and practices. Where the Spirit of the Lord is, there is freedom. For this reason, there is a contradiction of delivering a person that has already been delivered. This church is confusing the attributes of our spirit, with spirits of darkness. This is clear in the Spanish book; *"Teología*

Biblica y Sistemática" of Myer Pearman, that was published in English under the title: Knowing the Doctrines of the Bible in the year 1958 and translated to Spanish in 1992. This book is precisely one of the books that the Assembly Churches use in their Bible Institutes.

I quote textually: The Human Spirit

In every human lives a spirit given by God in individual manner (Numbers 16:22, 27:16). This spirit was formed by the Creator in the interior part of the human nature, and it is able of renovating and developing (Psalm 51:10). The spirit is the center and fountain of life in the man. The soul is the owner of this life and uses it, and through the body it expresses it. In the beginning God breathed life spirit into an unanimated body, and man became a living being. Therefore, the soul is a spirit that lives in a body or a human spirit that operates through the body, and the combination of both constitutes man in soul. The soul survives death because it is vitalized by the spirit, but both soul and spirit, are inseparable because the spirit is intertwined in the soul. They are infused amalgamated, in one substance.

The spirit is what sets the man apart from all things created and known. It contains human life and intelligence (Proverbs 20:27, Job 32:8), different from animal life. Animals have a soul (Genesis 1:20 in the original Hebrew) but not a spirit. In Ecclesiastes 3:21 it seems to refer to the beginning of life of the man and the beasts. Salomon formulated a question when he drifted away from God. The difference in animals is that they can't know Godly things (1 Corinthians 2:11, 14:2, Ephesians 1:17, 4:23) and they can't enter in personal relationships (John 4:23). When the spirit of man is inhabited by the spirit of God (Romans 8:16), converts into a worship center (John 4:23-24), songs, blessings, (1 Corinthians

14:15) and service (Romans 1:9, Philippians 1:27). The spirit, since it represents the most elevated nature of man, it is related to the quality of his character. For example, if they allow something to dominate them, it is said that they have a haughty spirit (Proverbs 16:18), according to the respective influences that control them, a man can have a perverse spirit (Isaiah 19:14), a provoking irritable spirit (Psalm 106:33), a precipitated spirit (Genesis 41:8), contrite and humiliated spirit (Isaiah 57:15, Matthew 5:3). Maybe they are under a servitude spirit (Romans 8:15) or impaired by a jealousy spirit (Numbers 5:14). They should protect the spirit (Malachi 2:15), dominate their spirit (Ezekiel 18:31) and trust God to change their spirit (Ezekiel 1:19).

When the bad passions dominate the man, and he manifests a perverse spirit, that means that the natural life or of the soul has dethroned the spirit. The spirit has fought and lost the battle. Man is prey of his natural senses and appetite. The spirit no longer exercises domain of the situation, and its lack of power is described like a state of death. From there it is necessary a new spirit (Ezekiel 18:31, Psalm 51:10) and only the one who breathed in the body of man the breath of life, can impart a new spiritual life. In other words, regenerate him. (John 3:8, John 20:22, Colossians 3:10). When this occurs the spirit of the man occupies a place of ascendance, and the man converts into a spirit. Regardless, the spirit can't live of itself, but has to seek a constant renovation through the Spirit of the Lord.

This information leaves it clear that almost all of the conditions of man, in the context I am using, they are related to our spirit, and nor spirits of darkness. It is also clear that all of these conditions go together with repent, through the constant seeking of renovation through the Spirit of God, and not through rebuking of spirits of darkness, that are generally not there.

Like I mentioned, there are churches that because of their beliefs and doctrines, put in danger the salvation of many souls. A clear example is the Catholic Church. The Catholic church teaches and preaches salvation through purgatory. After death, they pray rosaries to purify people, so they can present themselves before God since the bible teaches that nothing pure can be before him. And it is correct, nothing impure can go to heaven. What is incorrect about this doctrine is that there is a process established by God so that we can be purified. 1 Corinthians 15:51-52,

> *Listen, I tell you a mystery: We will not all sleep, but we will all be changed— 52 in a flash, in the twinkling of an eye, at the last trumpet. For the trumpet will sound, the dead will be raised imperishable, and we will be changed.*

> *Romans 8:17, Now if we are children, then we are heirs— heirs of God and co-heirs with Christ, if indeed we share in his sufferings in order that we may also share in his glory.*

Christ man himself, being perfect and without sin; also had to be glorified. John 7:39

> *By this he meant the Spirit, whom those who believed in him were later to receive. Up to that time the Spirit had not been given since Jesus had not yet been glorified.*

We will be purified in the resurrection when Christ resurrects us so that we face the judgment as per Revelations 20:11-15,

> *Then I saw a great white throne and him who was seated on it. The earth and the heavens fled from his presence, and there was no place for them. 12 And I saw the dead, great and small, standing before the throne, and books were opened. Another book was opened, which is the book*

> *of life. The dead were judged according to what they had done as recorded in the books. 13 The sea gave up the dead that were in it, and death and Hades gave up the dead that were in them, and each person was judged according to what they had done. 14 Then death and Hades were thrown into the lake of fire. The lake of fire is the second death. 15 Anyone whose name was not found written in the book of life was thrown into the lake of fire.*

That judgment is for everyone to be receive based on what they did while on earth. 2 Corinthians 5:10,

> *For we must all appear before the judgment seat of Christ, so that each of us may receive what is due us for the things done while in the body, whether good or bad.*
>
> *Hebrews 9:27, Just as people are destined to die once, and after that to face judgment,*

This proves that no prayer can change what has been established. The person that dies saved, is saved. The person that doesn't die saved, cannot be saved by anyone else. I don't want to be misunderstood; I am not saying that a Catholic can't be saved. What I am saying is that they can't be saved through purgatory. We have to fix things with God before we die.

I believe that in the beginning the Catholic Church adjusted itself to the word of God, (The Bible), but 300 years after its foundation, the Roman Emperor Constantino, did a reform in the church, and as Emperor made them include a big number of pagan beliefs. The list is long of the things that the Catholic Church practices and teaches, from the date of that reform until now.

There are churches that are subdivisions from the Pentecostals; dominate themselves prophets and apostles. This church started in the year 1928. As per its history, a Canadian man descendant from the Pentecostals, named Federico Mebius, arrived at El Salvador, CA, in 1927. He was the founder of this of this movement that later was taken to Honduras, Costa Rica, Mexico, and later to the United States. Now it is a world movement. This church, as well as the classic Pentecostals, and the neo-Pentecostals, have two identifiers, one part use the vail, the other half doesn't. As per theologians and connoisseurs of the doctrine, including Pentecostal pastors, evangelists, and even the rest of the religions, are being considered as false prophets due to the following information.

1) The prophets of the bible, the last one was Malachi, were directly commissioned by God or by an angel sent by God.

2) The prophets were infallible and authoritative; meaning you had to obey their prophecy, which was 100% without errors. (This was very important, because if they prophesized to you, and the prophecy didn't fulfill, it explains it all.). So, if they were prophets from God, the prophecy didn't fail.

3) The prophets could always prove their commission from God, they did miracles as supernatural proof provided by God.

What we see today in the apostles and prophets callings, are people auto denominated and mistaken, that can't do miracles or provide any evidence of their commission to prove and demonstrate that they are truly used by God.

A missionary in Africa, Gabriel Gil gave the following information:

This movement of the Apostles we have to pay close attention for the following reasons. The majority of the so called apostles, consider it an insult if they are not called apostle. They believe they have the monopoly of the anointing and coverage of God. They are the only ones that have the power to anoint other apostles. (The same practice as the Catholic Church). They believe they are bringing the kingdom of God here on earth.

Brother Gabriel says he has a document of the leaders of this organization that says: the apostolic government will annulated Satan's government. This is a clear attack against the rest of the Christians churches. They talk about restoring the apostolic posture, as if the Holy Ghost hasn't been working during over 2000 years here on earth. They believe that their apostle and his anointing are the ones that bring revelation to the church like the first days. Meaning they can bring a new doctrine. (The church needs to teach the doctrine that is already established in the sacred scriptures, and not establish a new doctrine). This means that these so called apostles, come with new revelations which the church needs to accept. This is very dangerous because by thinking they are the reformers of the church in general, means that the church that doesn't align to their doctrines, they end up being considered churches without anointing, without coverage, and the fullness of the Holy Ghost.

I used to sing with the Trio *Ecos Melódicos* for more than 10 years. We visited all the Christian Churches I have mentioned. We even went to sing at an event that the Catholic Church hosted for people with psychological problems. We visited several times the church of apostle and prophets. On all of the occasion there were prophecies from God as they alleged. I know people that go to this church, and they are always prophesizing about everything. This is what our Lord says in regard to these prophets. Matthew 7:22-23,

> *Many will say to me on that day, 'Lord, Lord, did we not prophesy in your name and in your name drive out demons and, in your name, perform many miracles?' 23 Then I will*

> *tell them plainly, 'I never knew you. Away from me, you*
> *evildoers!'*

The word of God has the complete revelation given by God. It contains everything that a Christians needs to know in regard to the spiritual, material, healing, present, future, and end, there is nothing that has to do with us that hasn't already been revealed in the scriptures.

God talks to women and men today, and gives them messages for the church, but everything is based on what is already established in the Bible. What has God established in His word?

Hebrews 1:1-2,

> *In the past God spoke to our ancestors through the*
> *prophets at many times and in various ways, 2 but in these*
> *last days he has spoken to us by his Son, whom he*
> *appointed heir of all things, and through whom also he*
> *made the universe.*

It is possible that in special cases God will give a prophecy but not in the way they are practicing and teaching it.

Everything that comes from spiritual means. Those words that state that God will give you house, a car, and all types of material prosperity I truly believe does not come from God. His word is clear, He knows what we need before we ask. A person that lives according to His word, God will provide them with everything they need, including the material things. 1 John 3:21-22,

> *Dear friends, if our hearts do not condemn us, we have*
> *confidence before God 22 and receive from him anything*
> *we ask, because we keep his commands and do what*
> *pleases him.*

A lot of people say that have asked, yet they have not received. We must understand that even though it says anything we ask, it also says that to receive we have to obey His commandments and do the things that are pleasant to Him.

> *Matthew 6:25-33, "Therefore I tell you, do not worry about your life, what you will eat or drink; or about your body, what you will wear. Is not life more than food, and the body more than clothes? 26 Look at the birds of the air; they do not sow or reap or store away in barns, and yet your heavenly Father feeds them. Are you not much more valuable than they? 27 Can any one of you by worrying add a single hour to your life[a]?*
>
> *28 "And why do you worry about clothes? See how the flowers of the field grow. They do not labor or spin. 29 Yet I tell you that not even Solomon in all his splendor was dressed like one of these. 30 If that is how God clothes the grass of the field, which is here today and tomorrow is thrown into the fire, will he not much more clothe you— you of little faith? 31 So do not worry, saying, 'What shall we eat?' or 'What shall we drink?' or 'What shall we wear?' 32 For the pagans run after all these things, and your heavenly Father knows that you need them. 33 But seek first his kingdom and his righteousness, and all these things will be given to you as well.*

Like I said, we don't need a new revelation from God, or for a prophet to reveal anything to us. Christ established in His word the way we would receive everything that we need, and everything is based on having faith. (everything that we ask for believing, we shall receive). Therefore, God has established the way we will receive everything we need. Now, the new prophets want to change what is already established, teaching that the way to receive is to "give money" to God; the more you give, the more you receive. Not in vain does the bible say that love for money is the root of all evil. Many people go to these churches that teach these false doctrines, because they want them to tell them what they want to hear; it's evident that they don't have knowledge of the word of God. What is established by God is for you to remain faithful and ask for everything with faith.

To participate in false doctrines is dangerous. God warns us not to participate of these false doctrines. Galatians 1:6-8 says,

> *I am astonished that you are so quickly deserting the one who called you to live in the grace of Christ and are turning to a different gospel— 7 which is really no gospel at all. Evidently some people are throwing you into confusion and are trying to pervert the gospel of Christ. 8 But even if we or an angel from heaven should preach a gospel other than the one, we preached to you, let them be under God's curse!*

Anathema is a Greek word; it is applied to a person that teaches a false doctrine; it means that they need to be excommunicated, separate them, or not allow them in the church of Christ, because they are cursed by God.

We have to be very careful. There are many religious leaders that are addicted with the thirst of power, prestige, and richness. And many Christians following them, and in many cases idolizing them, because they are famous.

Another thing that is difficult for me to believe in regard to this doctrine of apostle and prophets is, that it is not acceptable that the Lord waited until 1928, barely 90 years to establish a new doctrine, when the Holy Ghost has been ministering in the church sin Pentecostal for more than 2000 years. It's understandable the formation of the evangelicals, like I said for a bit longer than 500 years, the bible wasn't translated into other languages until precisely this time.

The theologies Mario E. Fumero says: It's enough already to see how people tithe the mint, the, albo, and the vegetable cumin of the poor, to satisfy the whims of the ones that exploit people in order to maintain as potentates the "apostles" that sell their faith, and the anointing's, to live as businessmen, with pride that is easily observed. They live grazing themselves and not imitating Jesus. All this at the expense of the unwary believers, that by ignoring the scriptures are victims of the exploitation.

The real reason of these behaviors we find them in Philippians 3:17-19,

> *Join together in following my example, brothers and sisters, and just as you have us as a model, keep your eyes on those who live as we do. 18 For, as I have often told you before and now tell you again even with tears, many live as enemies of the cross of Christ. 19 Their destiny is destruction, their god is their stomach, and their glory is in their shame. Their mind is set on earthly things.*

Of one thing I am sure, many churches have stopped being an organism, and have become an organization. They are more religious clubs. Week by week, month by month, year by year, they go to church and sing hymns, listen to a preaching, that generally are the same ones in various forms; and after many years they still have their houses founded in sand. Any wind blows and takes them away. In the majority of the churches, they no longer operate in the gifts of the Spirit, in Spiritual Tongues genuinely to the Spirit, because the vast majority of the people that are always speaking in tongues, knowing how to scrutinize the Spirits, they are human tongues, emotions and joy of the flesh. As evidence to this, a missionary that was invited to the *Tesalónica Church* in Bronx, NY over 40 years ago. She brought the following message on behalf of God to the church. She was telling them that while she was serving as missionary in India, one night God talked to her one night and said she would go to the United States, and you will give them the following message: *The majority of the churches are singing yes naya, for naya.* Confused, she didn't understand what this meant. She began to ask God for interpretation of the message. God spoke to her and said: when you go to the churches that I will take you, while they play music you will begin to see emotional people begin to speak in tongues that are not mine. As soon as the music stops, tongues disappear and all their emotions. Si naya. Satan's music. Music for Satan.

When a person receives the power of the Holy Ghost, it is impossible to act as nothing has happened a moment later. Another thing is that the Holy Ghost doesn't enter in anybody; there has to be purity and holiness. This of the baptism of the Holy Ghost and speaking in tongues, many people have taken it as something routine. (Very Dangerous) This means lying against the Holy Spirit.

In the majority of the churches there is no gifts of discernment, science, healing, miracles, tongues, interpretation of tongues, (nonintellectuals). The only gift that is used in the majority of churches today is, since we can't generalize, is prophecy since the one who preaches through the spirit, prophesizes. I know people that are old in the church and by their conversations and beliefs they don't have the true knowledge, of what His word talks about, spiritual knowledge.

Colossians 1:9,

> *For this reason, since the day we heard about you, we have not stopped praying for you. We continually ask God to fill you with the knowledge of his will through all the wisdom and understanding that the Spirit gives*

As I explained in all evangelical churches, they have different beliefs. We have been to Baptist Churches that the Pentecostals identify as cold and with solemn hymns. Nevertheless, there were other Baptist Churches that looked like Pentecostals. The same thing happened in Methodist Churches. On other occasions, in other churches we visited, members went outside to smoke and came back in to continue participating of the event. We were in a church in Queens, NY, where there were members of the church with lottery ticks in the middle of the church discussing the winning numbers. In a great number of Pentecostal churches, the service was a mess and with great irreverence. There were occasions that because of their style and beliefs, they sing and sing, and do everything, they finish at 11pm or 12 midnight, or even later, and say that the Holy Spirit took over the service. (The Holy Spirit I know is orderly).

In reference to what is happening today, the book of Matthew 24:12 says:

> *Because of the increase of wickedness, the love of most will grow cold,*

When we analyze the behavior of the majority of Christians today, we can say in all certainty, that we are currently living in these times. The truth is we don't really know who is Christian, and who isn't. The word says that by their fruits we will know them. Just because they go once or twice to the temple, and in some cases just on Sundays, week by week, and year by year, doesn't mean they are giving the results of the fruit the word talks to us about. This is just the same as a tree that has been planted for many years and has never given fruit. What is it good for?

Many churches seem not to be in tone with the word of God, because in reference to me, they come and go with the times and practice everything the world is doing. When I say many churches it's because we can't generalize, because there are still churches that are maintaining their light shining in the middle of darkness. When I talk about churches, I am referring to congregations, since the church of God like I said is one. But I believe that of these there are very little.

> *Matthew 5:13-16, "You are the salt of the earth. But if the salt loses its saltiness, how can it be made salty again? It is no longer good for anything, except to be thrown out and trampled underfoot.*
>
> *14 "You are the light of the world. A town built on a hill cannot be hidden. 15 Neither do people light a lamp and put it under a bowl. Instead, they put it on its stand, and it gives light to everyone in the house. 16 In the same way, let your light shine before others, that they may see your good deeds and glorify your Father in heaven.*

The church is the light of the world to dispel darkness of the immoral ignorance. It is the salt of the world to preserve it from moral corruption. I have been in church since I was born, and I have seen how the church has changed. For example: the way people dress, the worship and how it's been modernized, and in all senses, it has been transformed.

Famous theologians have taught us that since the first decade of the church, they conducted two types of services: one was of prayer, worship, and preaching; and the other was full worship known as party of love (Agape). To this party, only believers were able to assist. According to historians, in the first century spiritual songs were being written and sung with Psalms. It wasn't until like 30-40 years back, that all types of music were introduced.

Today we need to have more conscience of where we decide to congregate. Many ministers, evangelists, and prophets are teaching and practicing things that are not biblical. They are using the pulpits with the titles they have to take advantage of the members. They teach and even demand that the more money you give to the church, the more God will bless you. There are even some that dare to say: if you want a house, a car, whatever; give to God and He will give to you double. This isn't biblical, contrary, the word of God tells us the following:

> *2 Corinthians 9:7, Each of you should give what you have decided in your heart to give, not reluctantly or under compulsion, for God loves a cheerful giver.*

Let's analyze closely what it says: give as you proposed in your heart, not what they told you to give. Let us give not with sadness, but rather give with joy and happiness. And don't give just expecting for God to give something back in return.

> *Matthew 6:8, Do not be like them, for your Father knows what you need before you ask him.*

I believe it is offensive to God, to purposely give to him, just expecting something in return in material blessings. Imagine that you have a need for something, and your son or daughter for whom you have sacrificed all of your life, tell you: I will help you, but you have to pay me. A true man of God depends entirely on God. They don't have to use tricks for people to pay them.

> *1 Timothy 3:1-3, Here is a trustworthy saying: Whoever aspires to be an overseer desires a noble task. 2 Now the overseer is to be above reproach, faithful to his wife, temperate, self-controlled, respectable, hospitable, able to teach, 3 not given to drunkenness, not violent but gentle, not quarrelsome, not a lover of money.*

In God there are no investments like in Wall street. As Christians we must tithe and give offerings as much as we can with a grateful and happy heart. But, that business of giving to the lord expecting in return or of the more you give, the more blessings you will receive is a complete fraud and lie with lucrative purposes. As well as taking financial engagements without consulting with the Lord. This is why I have mentioned it is important to know the word of God. The person that doesn't have this knowledge is easily fooled with false doctrines and even worse, in some cases they are tricked on giving money to tricksters. Not to be misunderstood, to tithe is an obligation, and to offer is a blessing. Pastors are worthy of their salaries and the churches don't operate with air.

> *1 Timothy 5:18, For Scripture says, "Do not muzzle an ox while it is treading out the grain,"[a] and "The worker deserves his wages.*

Like I informed at the beginning, it is important to study where we will congregate in order to see if it is an organism or an organization. This explains why Jesus said in Matthew 20:16,

> *"So, the last will be first, and the first will be last."*

Many people buy jewelry that are supposed to be made of gold, but when they go to the jeweler, they find out that this shines like gold, but it isn't gold. It was just given a bath in gold, but inside it is not truly gold. In this same way there are Pastors, Evangelists and

Prophets that apparently shine as if they are women and men of God, but in reality, they are not. For someone that knows the word of God, it is easy to identify who has the calling of God and who doesn't. The ones that God calls don't lord the church. They are humble, compassionate, and fill all the requirements and ordinances given by God. When I see ministers and evangelists that seem prepotent, prideful, demanding, and even on occasions, dictators, that only speak on material prosperity and many times demand people to give them money; to me it's clear that they called themselves to the ministry. I say it like this people the person that is truly called by God, doesn't have this behavior. Like I said in the beginning, I base this on the word of God.

> *1 Peter 5:2-3, Be shepherds of God's flock that is under your care, watching over them—not because you must, but because you are willing, as God wants you to be not pursuing dishonest gain, but eager to serve; 3 not lording it over those entrusted to you, but being examples to the flock.*

As you can see, everything that shines isn't gold.

GOD CALLS US TO REPENTANCE

All of these behaviors are the warnings that our Lord Jesus Christ left us in His word, that would be happening, so that we could be alert and be vigil. I could understand that the people that have never known the word of God, don't understand the events that have already started taking place like the COVID-19 virus, the destructions by people with an uncontrollable and aggressive behavior, and even worse, backed up by corrupt politicians. The politicians believing, they are higher than God through their actions and words, like the case of those that think they can change the climate. The last thing they did was sign the peace pact that the bible tells us. When they say peace and security there will come massive destruction. There are pastors that don't believe and teach the opposite of the punishments already mentioned. The word will be fulfilled completely. While some Christians ignore that what is happening is the calling to repentance and the search of God, since the judgments that will take us to the end, have already started. God can't be fooled. You reap what you sow. At any moment the church can be lifted. 1 Thessalonians 4:13-17

> *Brothers and sisters, we do not want you to be uninformed about those who sleep in death, so that you do not grieve like the rest of mankind, who have no hope. 14 For we believe that Jesus died and rose again, and so we believe*

that God will bring with Jesus those who have fallen asleep in him. 15 According to the Lord's word, we tell you that we who are still alive, who are left until the coming of the Lord, will certainly not precede those who have fallen asleep. 16 For the Lord himself will come down from heaven, with a loud command, with the voice of the archangel and with the trumpet call of God, and the dead in Christ will rise first. 17 After that, we who are still alive and are left will be caught up together with them in the clouds to meet the Lord in the air. And so we will be with the Lord forever.

It's time to dominate our Spirit. Proverbs 16:32,

Better a patient person than a warrior, one with self-control than one who takes a city.

Not only dominate our spirit, but through repentance create a new spirit. Ezekiel 18:31

Rid yourselves of all the offenses you have committed and get a new heart and a new spirit. Why will you die, people of Israel?

The spirit, since it represents the most elevated nature of man, it is related to the quality of his character. For example, if they allow something to dominate them, it is said that they have a haughty spirit, according to the respective influences that control them, a man can have a perverse spirit, a provoking irritable spirit, a precipitated spirit, contrite and humiliated spirit. Isaiah 57:15,

For this is what the high and exalted One says— he who lives forever, whose name is holy: "I live in a high and holy place, but also with the one who is contrite and lowly

in spirit, to revive the spirit of the lowly and to revive the heart of the contrite.

The soul is sinful, so we shouldn't let it dominate our spirit. When the soul dominates your spirit, it means that your soul has defeated your spirit. This is why you have stopped being obedient to the divine amendments. The heart is the center of life of the desires, will, and judgment. Love, hate, determination, the good will, and joy. The heart knows, understands, deliberates, evaluates, and calculates. In the heart, thoughts are formed as well as the purposes, both good and bad. There is joy and pleasure in the heart. The heart is also the center of moral life. In the heart is where God has written his laws and where they are renovated through the operation of the Holy Spirit when we adopt the correct decisions. The correct thing to do is seek God in spirit and truth. Instead of criticizing the leaders that God in his omniscience has allowed being the Knower of all things; pray for them so that God can illuminate them. The Lord isn't seeking Priests with titles, or pastors with experience, or preachers with eloquence, or members singing I am leaving with Him. He is coming to lift a church without stains or wrinkles. That does not happen to us what happened to the ten fatwas virgins in Matthew 25.

To ignore the warnings can bring bad consequences. Clearly to many people will happen the same thing that happened to the people of Noah. Matthew 24:38-39,

> *For in the days before the flood, people were eating and drinking, marrying, and giving in marriage, up to the day Noah entered the ark; 39 and they knew nothing about what would happen until the flood came and took them all away. That is how it will be at the coming of the Son of Man.*

I say this because it is clear with the spiritual coldness people are taking the present events. Luke 21:9-20,

> *When you hear of wars and uprisings, do not be frightened. These things must happen first, but the end will not come right away."*

> *10 Then he said to them: "Nation will rise against nation, and kingdom against kingdom. 11 There will be great earthquakes, famines and pestilences in various places, and fearful events and great signs from heaven.*

> *12 "But before all this, they will seize you and persecute you. They will hand you over to synagogues and put you in prison, and you will be brought before kings and governors, and all on account of my name. 13 And so you will bear testimony to me. 14 But make up your mind not to worry beforehand how you will defend yourselves. 15 For I will give you words and wisdom that none of your adversaries will be able to resist or contradict. 16 You will be betrayed even by parents, brothers and sisters, relatives, and friends, and they will put some of you to death. 17 Everyone will hate you because of me. 18 But not a hair of your head will perish. 19 Stand firm, and you will win life.*

> *20 "When you see Jerusalem being surrounded by armies, you will know that its desolation is near.*

> *Jeremiah 6:16-19, This is what the Lord says: "Stand at the crossroads and look; ask for the ancient paths, ask where the good way is, and walk in it, and you will find rest for your souls. But you said, 'We will not walk in it.' 17 I appointed watchmen over you and said, 'Listen to the sound of the trumpet!' But you said, 'We will not listen.'*

18 Therefore hear, your nations; you who are witnesses, observe what will happen to them. 19 Hear, you earth: I am bringing disaster on this people, the fruit of their schemes, because they have not listened to my words and have rejected my law.

Jeremiah 7:21-28, 21 "'This is what the Lord Almighty, the God of Israel, says: Go ahead, add your burnt offerings to your other sacrifices and eat the meat yourselves! 22 For when I brought your ancestors out of Egypt and spoke to them, I did not just give them commands about burnt offerings and sacrifices, 23 but I gave them this command: Obey me, and I will be your God and you will be my people. Walk in obedience to all I command you, that it may go well with you. 24 But they did not listen or pay attention; instead, they followed the stubborn inclinations of their evil hearts. They went backward and not forward. 25 From the time your ancestors left Egypt until now, day after day, again and again I sent you my servants the prophets. 26 But they did not listen to me or pay attention. They were stiff-necked and did more evil than their ancestors.'

27 "When you tell them all this, they will not listen to you; when you call to them, they will not answer. 28 Therefore say to them, 'This is the nation that has not obeyed the Lord its God or responded to correction. Truth has perished; it has vanished from their lips.

Jeremiah 8:4-12, "Say to them, 'This is what the Lord says: "'When people fall down, do they not get up? When someone turns away, do they not return? Why then have these people turned away? Why does Jerusalem always turn away? They cling to deceit; they refuse to return. 6 I have listened attentively, but they do not say what is right. None of them repent of their wickedness, saying, "What have I done?" Each pursues their own course like a horse charging into battle. 7 Even the stork in the sky knows her appointed seasons, and the dove, the swift and the thrush observe the time of their migration. But my people do not know the requirements of the Lord. 8 "'How can you say, "We are wise, for we have the law of the Lord," when actually the lying pen of the scribes has handled it falsely? 9 The wise will be put to shame; they will be dismayed and trapped. Since they have rejected the word of the Lord, what kind of wisdom do they have? 10 Therefore I will give their wives to other men and their fields to new owners. From the least to the greatest, all are greedy for gain, prophets, and priests alike, all practice deceit. 11 They dress the wound of my people as though it were not serious. "Peace, peace," they say, when there is no peace. 12 Are they ashamed of their detestable conduct? No, they have no shame at all; they do not even know how to blush. So, they will fall among the fallen; they will be brought down when they are punished, says the Lord.

Jeremiah 9:1-11, Oh, that my head were a spring of water and my eyes a fountain of tears! I would weep day and night for the slain of my people. 2 Oh, that I had in the desert a lodging place for travelers, so that I might leave my people and go away from them; for they are all adulterers, a crowd of unfaithful people. 3 "They make ready their tongue like a bow, to shoot lies; it is not by truth that they triumph[b] in the land. They go from one sin to another; they do not acknowledge me," declares the Lord. 4 "Beware of your friends; do not trust anyone in your clan. For every one of them is a deceiver, and every friend a slanderer. 5 Friend deceives friend, and no one speaks the truth. They have taught their tongues to lie; they weary themselves with sinning. 6 You[d] live in the midst of deception; in their deceit they refuse to acknowledge me," declares the Lord.

7 Therefore this is what the Lord Almighty says: "See, I will refine and test them, for what else can I do because of the sin of my people? 8 Their tongue is a deadly arrow; it speaks deceitfully. With their mouths they all speak cordially to their neighbors, but in their hearts, they set traps for them. 9 Should I not punish them for this?" declares the Lord. "Should I not avenge myself on such a nation as this?" 10 I will weep and wail for the mountains and take up a lament concerning the wilderness grasslands. They are desolate and untraveled, and the lowing of cattle is not heard. The birds have all fled   and the animals are gone. 11 "I will make Jerusalem a heap of ruins, a haunt of jackals; and I will lay waste the towns of Judah so no one can live there."

Psalm 11:4-6, The Lord is in his holy temple; the Lord is on his heavenly throne. He observes everyone on earth; his eyes examine them. 5 The Lord examines the righteous, but the wicked, those who love violence, he hates with a passion.6 On the wicked he will rain fiery coals and burning sulfur; a scorching wind will be their lot.

Proverbs 1:24-26, But since you refuse to listen when I call and no one pays attention when I stretch out my hand, 25 since you disregard all my advice and do not accept my rebuke, 26 I in turn will laugh when disaster strikes you; I will mock when calamity overtakes you—

I am certain that the Holy Spirit has inspired me to write all of this information, from start to finish, for both Christians and non-Christians. It's on the reader to believe or not the warnings that God gives us. I save my responsibility in providing what God has given me.

The duty of the watchtower

Ezekiel 33,

> *The word of the Lord came to me: 2 "Son of man, speak to your people and say to them: 'When I bring the sword against a land, and the people of the land choose one of their men and make him their watchman, 3 and he sees the sword coming against the land and blows the trumpet to warn the people, 4 then if anyone hears the trumpet but does not heed the warning and the sword comes and takes their life, their blood will be on their own head. 5 Since they heard the sound of the trumpet but did not heed the warning, their blood will be on their own head. If they had heeded the warning, they would have saved themselves. 6 But if the watchman sees the sword coming and does not blow the trumpet to warn the people and the sword comes and takes someone's life, that person's life will be taken because of their sin, but I will hold the watchman accountable for their blood.' 7 "Son of man, I have made you a watchman for the people of Israel; so hear the word I speak and give them warning from me. 8 When I say to the wicked, 'You wicked person, you will surely die,' and you do not speak out to dissuade them from their ways, that wicked person will die for[a] their sin, and I will hold you accountable for their blood. 9 But if you do warn the wicked person to turn from their ways and they do not do*

so, they will die for their sin, though you yourself will be saved.

10 "Son of man, say to the Israelites, 'This is what you are saying: "Our offenses and sins weigh us down, and we are wasting away because of[b] them. How then can we live?"' 11 Say to them, 'As surely as I live, declares the Sovereign Lord, I take no pleasure in the death of the wicked, but rather that they turn from their ways and live. Turn! Turn from your evil ways! Why will you die, people of Israel?' 12 "Therefore, son of man, say to your people, 'If someone who is righteous disobeys, that person's former righteousness will count for nothing. And if someone who is wicked repents, that person's former wickedness will not bring condemnation. The righteous person who sins will not be allowed to live even though they were formerly righteous.' 13 If I tell a righteous person that they will surely live, but then they trust in their righteousness and do evil, none of the righteous things that person has done will be remembered; they will die for the evil they have done. 14 And if I say to a wicked person, 'You will surely die,' but they then turn away from their sin and do what is just and right— 15 if they give back what they took in pledge for a loan, return what they have stolen, follow the decrees that give life, and do no evil—that person will surely live; they will not die. 16 None of the sins that person has committed will be remembered against them. They have done what is just and right; they will surely live.

17 "Yet your people say, 'The way of the Lord is not just.' But it is their way that is not just. 18 If a righteous person turns from their righteousness and does evil, they will die for it. 19 And if a wicked person turns away from their wickedness and does what is just and right, they will live by doing so. 20 Yet you Israelites say, 'The way of the Lord is not just.' But I will judge each of you according to your own ways."

I only ask God to be the fountain of my inspiration, that what is written here can be of benefit to the readers, as I mentioned in the beginning, the results of ignorance will bring catastrophic results.

END

BIOGRAPHY

The author of this book, Obed Del Toro, was born in Cabo Rojo, Puerto Rico in the year 1944. Son of José Del Toro and Paula Lorenza Lugo (Loren). His roots and theological knowledge comes as a result of having parents that instructed him since his birth in theology studies. Both parents were theology teachers in the *Instituto Bíblico of the Asambleas de Dios in Bayamón Puerto Rico.* His mother graduated with high honors from the *Universidad Interamericana of Puerto Rico* college. She was a teacher in the public schools, missionary for 50 years at the council of God Pentecostal. Three of his five brothers have been ministers for over 50 years. Joel who has a doctorate in theology, was also the dean at *Instituto Bíblico de las Asambleas de Dios* in Bayamón, PR, was Treasury Secretary of the Hispanic. District of the West of the Assembly Of God, for nearly 15 years. His younger brother, Dr. José Del Toro Jr., has a doctorate in theology and is a phycologist. José works as a speaker in Latin America, the Caribbean, and United States since God has given him the blessing to have the theological and psychological knowledge and has been of blessing to many people in the countries mentioned. His sister Sara also graduated the *Instituto Bíblico of the Asambleas de Dios* in theology, and following the example of her mother, is also a minister in the Assembly of God, and has been preaching the word for over 40 years. She was a teacher at the *Instituto Bíblico of the Asambleas de*

Dios in Reading, PA and the Ft. Lauderdale, FL campus. His older brother Esteban and Obed were dedicated to the music and other ecclesiastic responsibilities. Esteban was cadet director at national level. Obed dedicated himself to music and to the study of biblical theology and systematic. In the year 1962 at the age of 18, he joined his brothers Esteban and José, and they formed a Trio of voices and guitars under the name *Voces Christianas*. This was a famous trio for religious music in the 60's in which they recorded popular songs such as: *"El Cojo de la Hermosa"*, and others composed by Obed, who inherited the music talent from his father and his mother who also wrote very popular hymns, heard and recorded by other trios and singers like: *"Las Espinas darán Rosas"*.

Obed later joined the brother Angel Pablo Ortiz and formed the *Trio Tesalónica* in Bronx, NY. In the year 1983 he joined Rubén Esquilín and Ángel P Ortiz to the famous *Trio Ecos Melódicos*. They recorded several productions of long duration in which the brother Johnny Colon participated and was a member of the trio, and the brother Edgar Siliezar, was the base player. As mentioned Obed is a composer and wrote music recorded mainly by *Ecos Melódicos* and *Voces Cristianas*, and other singers like Carmen Sanabria that recorded El Valle Del Dolor. At the beginning of the 90's, Obed moved to the state of Florida and started to work in the church as Minister of Worship. He also started to develop his knowledge in the Church Government in Biblical Theology and Systematic and helped write three constitutions at three churches. Obed is married to Antonieta Mirna del Toro of which three daughters were born: Loren Margarita, Xóchitl Garnet, and Valeria Antonieta. He is also the father of Obed Jr. Betzaida and Sandra.

I finish saying that I am a common citizen with very conservative convictions and based on the word of God. By nature, I have never been a person of parties or other activities. I have always been passionate and dedicated to studies of all materials that may benefit citizens both Christians or not.

Thank God I worked for 47 years straight without the need to be in an unemployment line. 25 years in the private business and 22 years for the Federal Government in the Postal Service. During these 47 years I had the privilege to serve as syndicate representative. 9 years in the private industry and 15 in the Postal Service of the United States, a total of 24 years as a representative and Defensor of workers with very good success.

I love God above all things, my wife, and my family, music since I was born, I love everyone, I love justice, I love the crib that saw me born Cabo Rojo, Puerto Rico, and the United States of America, with all of my heart.